Artificial Intelligence for Everyone

Steven Finlay

Relativistic

Relativistic

e-mail: AI@relativistic.co.uk

ISBN-13: 978-1-9993253-1-2

ISBN-10: 1-9993253-1-1

No plants or animals were mistreated in the writing of this book.

To Sam and Ruby

Contents

Acknowledgements

I would like to thank my family, and my wife Samantha in particular, for their support in writing this book.

Foreword

Artificial Intelligence (AI) is an important subject that I think everyone should know something about. Not least, because it's having an impact on everyone's lives every day – even if most people don't realize it.

However, most of the books I've come across seem to fall into one of two categories. The first are technical books, targeted at people with mathematics and IT degrees, who want to build artificial intelligence applications. The second type of book is aimed at the general reader, but to me, they seem to nearly always be overly evangelical in their approach. There is a lot of overhyping, too much emphasis on what *might* happen in the future and getting carried away by the propaganda put out by some of the tech giants of Silicon Valley.

There seems to be very few books that have tried to adopt a more realistic and pragmatic approach. In particular, books that are focused on the here and now, that seek to explain what Artificial Intelligence is, how it works and how it is being applied in the world today.

Hopefully, this is the book that will fill that gap.

Steven Finlay, January 2020.

1. Setting the Scene

"Americans worship technology. It's an inherent trait in the national zeitgeist."[1]

"Self-Learning Machines," "Algorithms," "Deep Neural Networks," "Robotics," "Automation." These are just a few of the terms being used to describe the seemingly endless torrent of new "Intelligent" tools, apps and gadgets that are sweeping across the world at an ever-increasing rate.

By all accounts, the big driver of these new technologies, "Artificial Intelligence," or "AI" as it is commonly abbreviated to, is already influencing or changing almost every aspect of our lives. This spans everything from how we work, travel and shop, to the way we obtain news and information, to the gadgets in our homes. It's also having a very significant influence on the human relationships we have with each other, how we communicate and how we express ourselves.

According to some, AI is awesome! Give it just another couple of years and no one will own or drive their own car. Google, Lyft or Uber's self-driving taxis will take us wherever we want to go. We won't need an army anymore because our borders will be secured by autonomous robot fighting machines. Medical apps, with a perfect bedside manner, will diagnose and even pre-empt our every medical need, delivering individually tailored treatments based on our DNA profiles and what our smart watches are reporting. Meanwhile, we'll all be living a life of leisure while a plethora of robots and digital assistants manage everything on our behalf. If you do happen to want a job, then a robot recruiter will interview you and make an impartial decision, with no preconceptions or unconscious biases about your age, gender, race or religious beliefs, as to whether you are right for the role or not. Utopia will have arrived!

But how much of this ultra-optimistic vision of a perfect machine dominated world will actually come to pass? Where and how quickly? What are the risks and dangers? Will it be a utopia or dystopia? Will we control the machines or will they control us? Does technology make us all more equal or exaggerate social inequality and isolation? What ethical questions do these technologies raise and how should society act to curtail their use? Is technology inherently racist and/or sexist? Could a robot become the leader of the Ku Klux Klan or another Hitler? What safeguards are required to stop this happening?

These are just a few of the questions that can arise when discussing these new AI-based technologies.

I have an opinion on all these questions, but unfortunately, I can't answer them for you definitively – no one can. Predicting the future is a tortuously difficult task. Even the best futurologists frequently get it wrong, and we should all be very cautious about believing anyone who claims they know precisely what will happen. New technologies are notoriously overhyped initially, even if they end up being really useful eventually[2].

A great example of this overhyping can be seen in the history of self-driving cars. For years, the tech companies were telling us that these things were just about ready to go. Every year, promises were made about self-driving cars being available by the end of the year, or early the following year, but each year the dates were pushed back[3].

What it looks like today, is that rather than the promise of owning a truly self-driving car that won't require a driving license to use, the near-term reality is self-driving taxis. These operate in well-defined "Geo-fenced" areas of a few major cities, overseen by human operators at taxi central. A self-driving car today doesn't mean owning something that sits in your garage at home that can take your children to school without you or whisk you across states on vacation. Sure, we are going to get there one day, but it's not a universal reality at the moment and is unlikely to be for quite a few years yet.

OK. So, it's impossible to predict the future with certainty. Also, it's important to remember that, unless you are of a fatalist persuasion, the future isn't fixed – it can be changed by our actions. We can all hold an opinion and draw our own conclusions. We can then campaign for or against what we believe, and for those who live in a democracy, vote for people who hold similar views to our own. It has been said that all forms of government are ultimately flawed, but most will listen to public opinion and will try to act in the interest of its citizens when new situations present themselves. We can see this principle in action with the backlash that has been seen against facial recognition software and the action that many governments and other jurisdictions are taking to ban or limit its use.

However, to be able to debate a topic in an educated and informed way, and hence come to a reasonable conclusion about it, you need an understanding of the thing being debated. AI is no different. To discuss it in a sensible way you need to understand the different types of artificial intelligence that exist and how AI works at a basic level (you don't need to be an expert). It makes sense to know what it is capable of and what its limitations are. What is real in the here and now? What's just around the corner? And what is pure speculation, about some distant future that your great-great-great-grandchildren might experience?

If your view of artificial intelligence is as some sort of magical techno-mysticism that does lots of clever stuff powered by teams of miniature unicorns, then who knows what nonsense you'll end up thinking. If I'm going to debate what the speed limit for cars should be in urban areas, then it makes sense that I know what a car is. I need to know what benefits cars deliver, roughly how fast they can go, the damage a car can do if it hits someone, plus some reliable information about the scale of the problem; such as the number of car accidents. Otherwise, hey, I could end up arguing that as long as the limit is no more than 2,500 miles an hour then no one is going to get hurt, right? What's your opinion on that? Do you agree?

Perhaps the most important thing to appreciate, and one of the biggest mistakes made by many tech workers and some futurologists, is to forget that technology does not operate in isolation. Society, morality, law, religion, business practices, human behaviour and our fears and prejudices all have a part to play. Technically, there's no reason why I shouldn't have my own private nuclear reactor installed in the basement to power my home. In reality, there are many practical reasons why I shouldn't have a nuclear power plant under my house and I don't think I need to go into the precise details as to why – it should be pretty obvious.

One of the key reasons why we don't have more driverless vehicles on the roads is that in the early days of their development not enough consideration was given to all the "Soft issues" that needed to be addressed. "If only we didn't have pedestrians and human drivers to worry about!" But, if anyone had given it any thought, then they should have realized, right at the outset, what the problems were. Banning people from crossing roads - crazy. The idea of Americans be forced to give up their right to drive? That would make removing the right to bear arms from the American Constitution look like a picnic. The developers of self-driving cars are now up to speed with these issues and accept that they have to deal with human habits, customs and behaviours, but it took a while for some of them to get there.

It's easy for someone coming from the tech sector to blame the politicians, and social and legal constraints as the reasons why their technologies don't become more widely adopted more quickly. Or sometimes, maybe they think it's all down to the irrational fears of the plebs at large, who should just put their fears aside and get on with things. But the principles that apply are no different to any other type of invention. AI tech doesn't have an opt-out to the normal rules of engagement. If I create a super-duper new type of smart phone, then I need to make sure all of the electrical components conform to the relevant health and safety legislation so that I don't fry consumers' brains. If I develop a great tasting new food additive, then let's get that fully tested before I start putting it into baby food and so on. I don't go about moaning that the law is an ass when it comes to these areas. I make sure I understand what the laws are and how they will affect me before rushing ahead with mass production of my product. In fact, I should be talking about these things in the very first design meeting, almost before I do anything else. Delivering a commercial product to the market is very much the last link in the chain.

The tech sector is learning but we can still see evidence of this narrow "Tech first" focused thinking in practice. For example, with the "Move fast and break things" philosophy that is popular with some Silicon Valley entrepreneurs[4]. Basically, get the technology stuff working and out to market as quickly as you can, and then worry about any problems later. Take crypto-currencies such as Bitcoin, and more recently, the Facebook backed Libra. Clearly, the inventors of these crypto-currencies didn't give enough consideration to the role and function of the regulatory authorities[5]. Consequently, there is considerable doubt as to how successful they will be in the future. The barriers to their wider use have nothing to do with the underlying "Blockchain" technology that Libra and Bitcoin are based upon. Rather, the biggest challenge is satisfying central banks and other authorities that they will not undermine the stability of the world's financial systems. Financial regulators have huge powers. They will only allow a new currency to become legal tender once all risks have been understood and mitigated against.

Without an appreciation of the context in which a technology is being applied it's also easy to fall into the trap of thinking that just because it's technologically possible it will happen. The fact that someone can build an intelligent fridge, that knows when you are running out of milk and reorders it for you automatically, does not mean that customers actually want that feature in a fridge. Or maybe they do, but not because they'll ever use it to buy milk, but because they can tell all their friends and neighbors about it. It's a status symbol rather than something that's practically useful. Goodness knows the number of drivers I've met who bought cars with self-park features who, after showing them off once, never used them again, or the mountains of virtual reality headsets that now lie idle.

As you can probably gather, my view is that the technology side of things is often the easy part. People and society are far more complex and difficult to deal with. Don't get me wrong, I'm not saying artificial intelligence tech isn't enormously complex, and it only exists because of some very clever people, but the fundamental elements of artificial intelligence systems are a lot easier to get to grips with than the nuances of the human condition.

With that in mind, the goal of this book is not to preach about AI and all the wonderful things that it *might* lead to. Nor is it my view as to what will happen in the future. Instead, the aim is to act as a guide to explain artificial intelligence and to put forward some of the arguments about its use and misuse. By understanding what AI is and how it works, this will help you form your own opinions as to how to approach artificial intelligence-based technologies. It's then up to you to decide what the pros and cons are and what, if anything, we should do about it.

With that in mind, the key things we are going to cover in the chapters that follow are:

- Discuss what artificial intelligence is and how it works.

- Understand where you are already likely to encounter AI in everyday life today.

- Describe how artificial intelligence can be used to the benefit of individuals, business and society.

- Describe the risks and dangers, legal and ethical issues around the use and misuse of artificial intelligence.

- Mull over some of the things that may (or may not) be on the horizon in the next few years.

As someone who has become a little hacked-off with a lot of books that seem to be to written by people who spend their time worshipping at the techno-altars of Silicon Valley, I think I should also be very clear as to what this book is not.

- It's not a book for Geeks or Nerds (there a difference![6]) No technical stuff here my friend. All explanations are simple, with no fancy Greek letters that those guys call math. There is a very tiny bit of arithmetic in the book somewhere, but it's very much of the $2+2 = 4$ kind. Not the $E=MC^2$ variety.

- It's not a pro-AI propaganda piece. Everything is not 100% awesome in the world of AI. Sure, it's really interesting, is changing society in all sorts of ways and has some great applications, but it's not a universal panacea for all the world's ills and may even create some new ones.

- It's not an anti-AI piece either. I've seen what the stone age was like and I don't want to go there. I want to move forward not backwards. Something doesn't have to be perfect to be of benefit, it just needs to be better than what came before. Where, on balance, AI can make life better, let's get on with it.

- It's not written by someone employed by a tech giant or other interested parties. The author, has worked in AI related fields in the past, but regards himself as pretty impartial these days. He can also confirm that he has not been paid by the Russians or North Koreans to corrupt your thinking about American technology companies.

In terms of book structure, then it's pretty much a game of two halves. The next four chapters are very much about understanding AI, what it is and how it works. The final four chapters are more about how AI impacts what we do and the legal and ethical issues that it raises.

Oh, and before I forget, if you see any text in ***bold italics***, then that means it has an entry in the glossary at the back of the book.

Fantastic. Now that we've got that sorted, let's move on to the next chapter and chew over what artificial intelligence actually is.

2. What is Artificial Intelligence?

"Any artificial intelligence smart enough to pass a Turing test is smart enough to know to fail it"[7]

OK. So, I hear this term "Artificial Intelligence" being talked about all the time, but what is it? It's supposed to be everywhere but I haven't yet met an **android** that I can have a meaningful conversation with. I keep hoping to meet one in a bar or at my next book club meeting, but no luck yet.

There are loads of different and varied explanations as to what artificial intelligence is but, I'm afraid to say, there isn't one universal definition that everyone agrees with. Not to worry. Let's have a bit of a poke around and see what we can come up with.

A simple and very practical explanation is that something human-made, a machine, that can act and reason just like a person embodies all that it means to be artificially intelligent. This isn't an unreasonable place to start and leads one to envisage a world filled with human-like robots doing human-like things. Possibly our servants, conceivably our masters, maybe our friends. But what about dogs and dolphins? Don't they display a degree of intelligence also? Mmm. Let's leave animal intelligence out of the debate for now.

If we were to go back in time a few years, then this description of human-like entities behaving in a similar way to ourselves is one that many people would have readily agreed with without much further thought. In fact, the ability to replicate human behaviour is a key feature of the famous "Turing test." This was devised by the mathematician Alan Turing,[8] back in the mid-20th century, to determine if a computer can be considered intelligent.

In the test, Turing envisaged a human in one room and a machine in another. A human judge then converses with the human and the machine remotely, without knowing which is which. The judge may engage them in conversations about the weather, politics, the latest clip they'd seen on YouTube or anything else they liked. The machine is deemed to be intelligent if, after a period of time spent in conversation, the Judge is unable to distinguish between the human and the machine.

At the time of writing no machine has passed the Turing test, although there have been some very good efforts, and it will probably be many years before a computer is able to do so convincingly and repeatedly[9].

The Turing test is pretty cool, but it's not without its problems. Lots of people have tried to write computer programs that mimic human responses in an attempt to fool the judge and pass the test, and when the test is finally passed, this is probably the route that will be taken. The focus has been on passing the test, rather than creating something that is actually intelligent; i.e. clever programming tricks designed to deceive are not in the spirit in which the test was originally conceived. To put it another way, these people who are trying to pass the test haven't built a really kick-ass robot, and then said: "Oh, by-the-way, I wonder if it would be any good at the Turing test? What do you think Robot, do you want to give it a go?" Rather, they've said "Mmm, let's see, what do we need to do to pass the Turing test? and let's try and build something that does that."

Another argument against using the Turing test to assess intelligence is that a machine doesn't need to be conscious in the way people are to pass the test. It doesn't need to be self-aware of its actions or have any understanding of the responses it makes. It can pass by just blindly following some (very complex) computer code.

If we stick to the concept of a fully conscious self-aware thinking machine, with all the different mental capabilities that we have, as our definition of artificial intelligence, then we are years, if not decades, away from such machines. In fact, there is an argument (that many scientists dispute) that no system, based on the types of computers we use today, could ever be truly conscious as we understand it, no matter how powerful they are. There must be some additional (as yet undiscovered) element required for human-like consciousness and self-awareness that can't be replicated via computation alone[10].

Even if we remove the requirement to be conscious and self-aware, and instead focus on the ability to mimic the full spectrum of human intellectual capabilities using massively powerful computers, then there is nothing anything like this in the world today. This broad "Do it all" type of artificial intelligence, or ***General AI*** as it is often referred to, is also likely to remain in the realms of science fiction for some time to come.

Taking a slightly more philosophical line, isn't it a little arrogant to define intelligence as a solely human property? Could there not be other, completely non-human (alien) ways of thinking that can also be classified as intelligent? A machine that thought in this "Other way" would therefore be intelligent, but possibly not in a way that we could readily comprehend.

So, fully human-like AI is still a distant dream. But don't be too disappointed. Instead, let's lower the bar a little. Instead of thinking of AI as something that must be like a person (or an alien), let's adopt a less stringent, and somewhat narrower, definition of artificial intelligence. If a machine can reason, such that it can make decisions that are at least comparable to those that a human would make, then that's artificially intelligent behavior. To put it another way, a machine that can gather information, learn about a situation and then decide upon a sensible course of action based on what it has inferred is behaving intelligently.

If we want to split a few hairs, then saying that something is "Behaving intelligently" may not be quite the same thing as "Being intelligent" but let's assume that that's good enough for now. More concisely, we can summarize this somewhat restricted definition of artificial intelligence as follows:

> Artificial Intelligence (AI) is the ability of a machine to assess a situation and then make an *informed decision* in pursuit of some *aim* or *objective*.

This definition encompasses the replication of both conscious and unconscious decision-making in humans and other animals. An example of conscious human decision-making is where someone is assessing the experience of different job candidates with the aim of deciding who is best suited to fill a vacant role. A machine that could do this instead of a hiring manager would qualify as being intelligent within this area of expertise. An example of unconscious decision-making is taking in the features of a picture before deciding that you are looking at a cat rather than a cake. Again, a machine that can do this would fall within our definition of artificial intelligence. Both of these are examples of artificial intelligence that are in widespread use today.

In displaying intelligent behavior, a machine can assess the information it has to hand and make a decision on the basis of what it knows – an *informed decision*. If things change; i.e. new and different information is made available, then the machine will reassess the situation and a different decision may result. For example, introduce pictures of potatoes, and the object recognition system can learn to identify not just cats and cakes but potatoes as well.

If we look at the landscape today, then there are lots of clever systems that meet this definition of artificial intelligence. When you here about some new application of artificial intelligence, this is pretty much the definition that is being referred to. Now, I'm not saying that the definition of AI that I've just presented is one that all the experts will agree with or is theoretically complete. In fact, it would probably have many of the original pioneers turning in their graves. However, from a practical perspective, it pretty much covers all of the tools, apps and gadgets in the world today that the tech companies describe as incorporating artificial intelligence.

Given the on-going hype around artificial intelligence, what you tend to see is that organizations with products that incorporate even the tiniest bit of learning and adaptation will describe these product as "Incorporating AI" – even if it means stretching the truth a little. This is to ensure that they sound hip, down with the influencers and at the cutting edge of innovation. A few examples I came across that seemed to fall into this category are:

- Colgate's Connect E1 Smart Electronic Toothbrush with "Intelligence embedded in the handle."[11]

- Samsung's "AI-powered intelligent laundry assistant."[12]

- L'Oréal's foundation make up matching machine which promises to find the "Exact match for your skin using AI." [13]

OK – so where is this type of limited AI being employed today? All over the place, but a few examples of where you are likely to encounter AI powered applications in everyday life are:

- **Content detection.** The aim of the system is to identify if text, pictures or speech relates to certain subjects. These systems are used by social media companies, such as Facebook, to identify and remove offensive and/or illegal material posted to their sites so that you don't see them.

- **Chatbots.** The objective of a chatbot is to provide answers to peoples' questions. Simple chatbots are routinely used by all sorts of companies to automatically answer basic customer queries. More complex examples of chatbots are personal digital assistants such as Amazon's Alexa, Apple's Siri and Google's Assistant.

- **Streaming service recommendations.** Content providers, such as Netflix and Spotify, use AI tools to learn the preferences of individual users and hence the best song to play next or what programs to recommend.

- **Language Translation.** Tools, such as Google Translate, are AI-driven apps that can translate almost any language into any other.

- **Fraud detection systems.** The goal of these systems is to identify illegal transactions. If you get a call from your bank querying a credit card purchase, then it will be because an AI-based system has identified that transaction as suspicious.

- **Home thermostats.** Systems such as Google's Nest are relatively simple examples of AI that manage a home's heating system. The aim is to optimize the temperature for the owner while at the same time minimizing energy usage.

- **Policing and law enforcement.** AI is used for many purposes. One example is making parole decisions. The goal is to only recommend parole for prisoners with a low risk of reoffending. A second example is facial recognition to spot criminal suspects. Another is the use of AI to help decide which areas police should patrol to maximize arrest rates.

- **Dating compatibility.** The system examines the characteristics of different people with the goal of deciding who is most likely to be compatible with who.

- **Financial products.** Anytime you take out a loan, ask for an insurance quote or get a phone contract, AI-based systems will be making the decisions in almost every case. People aren't involved in the decision-making process at all.

With the exception of personal assistants such as Siri, Alexa, etc., most of the AI you are likely to encounter will be occurring behind the scenes. There isn't an obvious artificially intelligent agent at work, who looks you in the eye, engages you in meaningful conversion and then does something. Most people probably don't even know when AI is being used. Things just happen, and usually happen better than they used too, because of the AI tech that's being employed. That's either reassuring or very disturbing, depending on what the AI is being used to control and how the actions it decides impacts you as an independent, self-determining being. Maybe there should be a "Warning on the packet" whenever AI is being used to do something that might affect you.

If you want to see some modern "Pure AI" in action, then Microsoft's "CaptionBot" is an interesting place to start[14]. Just upload a picture to the website and the AI software will describe what it sees. I found it quite fun because although the bot describes some things quite well, it sometimes gets things wrong or gives vague / non-sensical answers. A picture I showed it of someone clearly smoking a cigarette was described as: "A man brushing his teeth" and an image of a woman juggling prompted the reply: "I think it's a large white ball." – well, not a million miles away and quite impressive, but as it stands, I'd put my money on your average three year old to provide more informative descriptions.

Another one to try is MuseNet[15]. This uses AI technology to generate completely new pieces of music once provided with a few preliminary notes to get it started. A third example, that I found slightly spookier, is the: thispersondoesnotexist[16] website. This generates pictures of people that look very real – but are completely artificial. The people in the photos don't exist! I couldn't tell that the people weren't real. It was only after looking at many different artificial pictures of people that I decided that maybe the eyes didn't seem quite right to me. However, I suspect that that may just be my unconscious bias at work because I wanted to be able to say that I could spot the difference between a real photo and a fake one, even if I can't.

If you want to see some great examples of artificial intelligence combined with advanced robotics, then Boston Dynamics has a lot of videos of their intelligent robots in action available to view from their website[17].

Moving on to think about some examples of things that are being trialed, or have been showcased at technology forums, and which might become mainstream in the very near future include:

- **Fully self-driving cars and trucks.** Self-driving vehicles combine a number of technologies, but AI forms a key component of the overall package. In particular, the AI provides the decision-making logic that aims to get you to where you want to go, while keeping within the law and avoiding accidents.

- **Fruit picking robots.** Picking fruit is a more complex task than you might think. AI is being incorporated into robots to guide them to perfectly ripe, non-rotten fruit, and then to pick it without damaging the parent plant.

- **Personalized in-store advertising.** A combination of AI technologies, including facial recognition that identifies you as you enter a store, software that can determine your mood and generate predictions about your buying preferences are used to deliver individually tailored ads as you move around the store.

- **Identifying potential best sellers.** AI can be used to identify which songs, books or films are most likely to be successful. Shortlists of promising artists and writers are then passed on to human editors to make the final decision as to which to take forward.

- **Predicting the outcome of criminal trials**. Legal documents and evidence are reviewed with the aim of deciding if a case is likely to be successful. Only cases that the AI predict have a good chance of leading to a prosecution are taken forward to a full trial. The use of automated AI-based judgements is also being trialed in civil cases and small claims courts[18].

- **Autonomous weapons.** Adding a grenade and facial recognition to a drone isn't difficult. Putting a more sophisticated "Seek and destroy" capability into a piece of military hardware isn't much harder. The reason why this type of weaponry is not more prevalent has more to do with moral/political constraints than technological ones.

One of the key trends we are now seeing are applications of AI that are moving beyond just assessing information and decision-making, and more into the physical world; i.e. combining the analytical and decision-making abilities of artificial intelligence with advances in engineering and robotics. This facilitates the automation of many physical tasks that would once have been undertaken or controlled by a person. Large modern warehouses are a prime example of a working environment that has seen a proliferation of robots in recent years, with far less requirement for human staff to do things like stocking warehouse shelves or picking items for use/dispatch.

A key feature of all the aforementioned applications of artificial intelligence is that they are examples of what is termed "***Weak AI***" or "***Narrow AI.***" The systems that do these things can be extraordinarily good at doing certain tasks, often far better and faster than the best human, but that is all they do. They can only make informed decisions about a very narrow range of problems that they have been designed to deal with. If you want to use AI tech to do something else, then you have to develop a new app from scratch.

Consider an object recognition system, such as one that might be used in the very important task of differentiating between cats, cakes and potatoes. The goal of the system is to correctly identify what object the camera is looking at. The AI has no other capability or objective other than recognizing objects. It can't, for example, decide who is likely to be a good match for you on a dating site.

Similarly, consider a cutting-edge medical AI application that can assess a patient's symptoms and come up with a diagnosis just like a GP can. It may be excellent at determining if there is anything wrong with you, but it won't have a clue as what your insurance premiums should be or whether you are interested in the latest 2-for-1 deal at your local supermarket. Its range of application is limited to medical diagnosis. A tool such as Google Translate is fantastic at what it does, but will never be any use at detecting credit card fraud or deciding who's the best candidate for a job.

Another weakness of many of the current state-of-the-art AI systems is that they are able to do some extraordinarily clever things and seem very clever when you focus them on very specific tasks, but they can also behave in unbelievably stupid ways if you take them just a short way outside of their comfort zone. As one pioneering AI expert commented[19]:

> "So many people overestimate the intelligence of these systems. AI's are really dumb. They don't understand the world. They don't understand humans. They don't even have the intelligence of a 6-month-old" – Yoshua Bengio

In formal scientific studies, today's best object recognition systems are now better at identifying normal everyday objects in pictures than people are. However, engaging in a little mundane trickery, such as turning pictures through 90 degrees, can cause these systems to struggle[20]. Any normal person can recognize a bus that has rolled on to its side, but many AI systems can't – which is important for things like self-driving vehicles if they need to be able to detect accidents.

Likewise, changing just a few pixels in an image, making no discernable changes to the image that a person would notice, is enough to confuse them[21].

Objects that contain images of things can also cause problems that humans wouldn't blink an eye at. If you buy one of those birthday cakes that has edible photo icing with a picture of your cat on the top of the cake, then the system will, more likely than not, report that it sees a cat rather than a cake.

Chatbots provide another great example of the limitations of current AI capability. Perhaps the highest profile "Fail" was Microsoft's Tay chatbot. Tay was developed by Microsoft with the intention of talking to teenagers on-line via Twitter. As it engaged in more conversations so it would adapt and contribute in new and different ways.

Tay's developers seemed to not have considered the nature of some teenagers or some of the nastier elements of social media. Tay ended up being withdrawn within hours because it started to tweet sexist, neo-Nazi and other inappropriate content[22]. The designers of Tay didn't intend this, but all Tay did was to learn and adapt to the pattern of conversations it was exposed to. So, when people started making inappropriate comments, Tay's programming assumed these were normal and played them back in its own responses.

Tay's ability to respond using natural language could be considered almost human, but its inability to moderate its responses to cater for the social and ethical norms society takes for granted was very sub-human.

A valuable lesson to learn from Tay is that it also demonstrates another, and possibly one of the most serious, limitations of current AI. Artificial intelligence based tools have no social awareness, no moral sense and no conscience. They just act to achieve what they have been tasked to do. Nothing more and nothing less. There is no angel sitting on their shoulder to provide them with moral guidance.

Narrow AI is currently where we are at, at the start of the third decade of the third millennium. However, where software developers have been clever is that they have realized that they can create some really wonderful apps if they combine several types of Narrow AI together. They can then add some human defined common sense rules to deliver a tailored and far more expansive user experience. If you take something like Amazon Alexa, then there are a whole host of separate systems that have been brought together to do several different things in an integrated way.

With a typical digital assistant, there is one type of AI software that works out what you are saying, another that can search the internet to find the information you are seeking, another to translate that into something meaningful that can be read back to you and so on. Where the system has been seen to generate stupid or incorrect answers in the past, then human programmers will add rules to override what the AI would have responded with.

Humans are still very much in the loop when it comes to making these systems better. A company like Amazon employs a lot of people to listen to samples of what people have said to Alexa to help improve it[23]. Humans will also program responses to certain types of questions. If you ask a digital assistant to "Tell me a joke" then a joke will be selected at random from a human complied list and played back to you. It won't be because the digital assistant has learnt jokes directly from its conversations with users. The overall effect is to deliver a service that is better than either artificial intelligence or human intelligence could deliver on their own.

The same principles apply with complex robotic systems and autonomous vehicles. One type of AI is used to visually identify objects and features, another to interpret voice commands, another to determine how to navigate the environment and so on. Humans then add rules to deal with any specific conditions that the AI elements can't deal with adequately.

Despite their current limitations, these types of tools and services are evolving and improving all the time. For example, Amazon has developed a system to allow software developers to create new applications and services that can be integrated with Alexa to expand the range of what it can do[24]. It's not beyond imagination to see a time, in the not too distant future, where we end up with something that delivers a broad human-like experience, that is not a million miles away from what people envisage as General AI, but is delivered simply by having thousands of narrow AI applications seamlessly integrated together to be able to undertake almost any task imaginable. Currently, Amazon claim that there are more than 100,000 different apps supporting Alexa to enable it to do all the different things it is now capable of[25].

There has been an explosion in the applications of artificial intelligence in the last few years, but what may be surprising is that real world applications of AI technology aren't something new. Some have been in widespread use since the mid-twentieth century.

What may have been the first commercial use of an AI technology was in credit granting. Back in the 1950s, banks and other providers of consumer credit used to employ large teams of highly trained underwriters to decide who to grant credit too. They would weigh-up all the information they knew about someone and then decide if lending money to them represented a good risk. Every loan was manually assessed. Being an underwriter was an important and well paid job, and it would typically take several years to rise to the top of this particular profession.

When commercial computers become available, lenders realized that they could use this new technology to analyze details of thousands of historical loans to identify the features of customers who tended to repay their loans and the features of those who were likely to default. All these different features could then be combined together to calculate a probability of default for anyone, given someone's unique set of characteristics and circumstances. For example, their age, income, type of job, marital status, where they lived and so on. These probabilities of loan default were branded "Credit scores." It was then a simple task to automate the calculation of credit scores for each new loan application, and to set rules to decide who to grant credit to on the basis of the score that they received.

As lenders processed more loans and gathered more information, so the calculation of credit scores were continually refined and improved[26].

What the banks and other lenders soon realized is that these automated underwriting systems could outperform the best human underwriters. Their use resulted in more good loans being granted and less bad ones. But that wasn't the end of the story. A human typically took several minutes to assess each loan application and make a decision. With automated credit scoring, it only took a fraction of a second. This meant that the machines also offered a much faster and cheaper way of making lending decisions. The writing was on the wall for the underwriting profession.

These days, automated credit scoring rules the roost. Human underwriters do still exist, but they are a rare breed these days, who are only used to assess loan applications in a very limited set of circumstances. One situation where they are still required is where a dispute arises and the law requires human intervention. Alternatively, there are sometimes unusual or one-off cases that the automated systems are not programmed to deal with and so the case gets referred to a real person to consider.

Today, automated credit scores are not just used to make decisions about loan applications or other types of consumer credit. They are also used to support all sorts of other financial decisions. In the world of insurance, the evidence suggests that uncreditworthy people; that is, those with the lowest credit scores, tend to make the most insurance claims. Therefore, these people tend to charged higher insurance premiums than everyone else[27]. Landlords will often think twice about renting their property to anyone whose credit record is below par. Likewise, some companies view a good credit score as a measure of an employee's probity and will often reject an applicant whose credit score is too low. Your credit score is a very significant thing, that can impact your life in all sorts of different ways.

Credit scoring and other automated decision-making systems that existed back in the 1950s and 60s were very simple by today's standards. A modern AI practitioner (a **data scientist**) might laugh if you were to suggest that these systems constituted what they would call artificial intelligence. However, the idea of learning from large amounts of data, predicting outcomes and then making decisions on the basis of those predictions – all in an automated way, are in principle no different from the vast majority of AI applications in use today.

What has changed in recent times is the complexity of the systems that are now available. This is partly due to the "***Big Data***" world we now inhabit. It's easy to forget that before the early 2000s very little personal data about us was stored electronically. Organizations may have known about your financial status each month end and if you paid your bills on time by obtaining data from a credit reference agency, but very little else.

With the advent of online shopping, smartphones and social media in particular, there was an explosion in the amount of detailed personal information that was available. Access to this information has been vital to allow AI systems to learn about us and our activities. In particular, to learn how our data trail can be used to predict our future actions, and hence, how best to deal with our anticipated behaviours. Similarly, there is now massive amounts of information about all sorts of business and industrial processes. This is being used to support the development of AI-based tools that are helping to optimize these processes in businesses and factories around the world.

The second element that has fueled the rise of AI is the vast increase in cheap data storage and the computational power of modern computers, under-pinned by some very clever algorithms (computer code). Much of the theory behind modern AI systems was developed many years ago back in the 20th Century, but has only become feasible to deploy with the availability of massive amounts of cheap computing power. Combine vast amounts of data with very capable computer hardware and sophisticated software, and you can build machines that do all sorts of clever things.

Things have also been helped along immensely by the democratization of many artificial intelligence technologies. Once, access to the technology required to develop AI solutions existed only in academia or the research divisions of the largest companies. If you wanted to play with AI software, then it would cost you thousands of dollars to buy it. These days, there are a whole host of (mostly free) software tools available to allow people to develop AI applications on their laptop or home PC, or by using cloud services offered by the IT giants such as Microsoft[28] and Google[29]. If a college student or home enthusiast wants to include voice recognition in the app they are developing, then they can just use the free version of the Alexa software that Amazon has made available[30].

Things are also helped by the fact most of the tricky mathsy stuff has been automated. You don't actually need to understand much of the maths to be able to apply cutting-edge technology. In some ways it's just like a modern car. You can be a very good professional driver, yet have almost no understanding of what goes on under the bonnet.

What this means is that the skills required to use AI software, to build something intelligent, doesn't require you to be a Nobel prize winner or mathsy PhD. A typical college graduate, who is numerate and with some computer programming skills, can start producing AI apps very quickly. Sure, we are not talking about them being able to develop a self-driving car in a few days, but for a simple AI-based app, such as a basic chatbot, then a little initiative and some basic knowledge of computer programming will get you a long way very quickly.

When talking about artificial intelligence, you will almost invariably come across the term *Machine Learning*. What's that? OK – let's go a little further down the rabbit hole and try to explain what machine learning is. In a sentence:

> Machine learning is the use of mathematical procedures to analyze and infer things from data.

The goal of machine learning is to discover useful things about the relationships between different bits of information. It identifies patterns in data. In particular, correlations between one thing and another. Some examples of individual correlations are:

- The number of children born in Spring is higher than the number born in Autumn.

- High income households tend to consume more beef than lower income households.

- Asthma is more prevalent amongst city dwellers than country folk.

- Men who buy red or yellow shirts are more inclined to buy black shoes, but men who buy green shirts tend to buy brown shoes.

- People who earn more tend to live longer than those who earn less.

An important thing to appreciate is the difference between **correlation** and **causation**. We are not saying that Spring causes more births or that having a high income results in you eating more beef. All we are saying is the when we see a change in one thing, we also see a change in another. The fact that two things are correlated does not necessarily mean that one causes the other to happen.

Once the relationships (patterns or correlations) have been identified, then these relationships can be used to make inferences about the behavior of new cases when they present themselves. The more information that is made available, the more accurate and nuanced the inferences about the various relationships becomes. To put it another way, the more data you feed machine learning with, the better it becomes at figuring out what is going on.

In essence, this is analogous to the way you and I learn. We gather information by observing what goes on around us and draw conclusions from our experiences about how the world works. We then apply what we have learnt to help us deal with new situations that we find ourselves in. The more we experience and learn, the better our ability to make decisions becomes.

Where machine learning comes into its own is where there are large numbers of subtle correlations between lots of different, sometimes seemingly unrelated, things and a particular outcome or behaviour. A doctor may be able to weigh up maybe five or six different factors when coming to a decision about a diagnosis. This compares with a machine learning based system that can consider thousands of correlations between different data items and different medical conditions all at the same time. This is a very powerful thing to be able to do. As at the time of writing, AI systems are often better than human experts at diagnosing specific conditions, and it's probably only a matter of time before their abilities equal or exceed those of your average GP.

An absolutely fundamental application of machine learning is prediction. In fact, you could almost say that machine learning *is* prediction. Nearly all applications of machine learning aim to deliver a system that allows you to determine something that you don't know, based on the information that you currently have available. Given what I know about what has happened in past situations, what is most likely to occur in this new situation in which I find myself? An appropriate decision can then be taken based on the outcome that is believed to be most likely to occur.

When machine learning is applied to the vast quantities of personal data that is held about us, then this translates into systems that can predict how people are expected to behave in the future based on information known about them today. Having identified the relationships that exist, and given detailed information about a person, it is possible to make very accurate predictions about their future behavior based on what they did historically, combined with their situation today.

A machine that can predict what you are going to do? – scary huh? Yes, but not half as scary as a government or corporation that has control of that machine and what it may try and make you do on the basis of those predictions.

Direct marketing is probably the area where machine learning has been most widely applied. If you give me a sample of peoples' data and previous purchasing history, I can utilize machine learning to identify patterns in purchasing behavior. These types of people buy this type of product, those types of people buy that type of product. I can then use these patterns to predict what goods someone is likely to buy next; i.e. future purchases are the outcome that I want to predict.

When a company such as Facebook or YouTube uses an AI-driven marketing app to decide what ads to present to you, this is based on the historic analysis of millions of people's characteristics such as their gender, where they live, their browsing history, the content of their posts on social media and what they subsequently bought. This model of purchasing behaviour can then be applied to you and your set of characteristics now, at this precise moment, to predict what you are most likely to buy next, and hence, what ads you should see. As things about you change over time, the predictions are updated and consequently a different set of products may be presented to you.

If we return to our discussion about credit scoring, then things are very similar. Loan companies apply machine learning to historic loan agreements to identify features of good paying customers. It may find, for example, that older people tend to better at paying back their loans than young people, or accountants tend to be worse than school teachers. Once the relationships in the data have been identified, these can be used to predict how likely a new credit applicant is to repay their loan when they apply. Your credit score is, to all intents and purposes, a prediction as to how likely you are to repay your debts in the future.

A fundamental feature of this process, which is very important to understand, is that the machine learning process considers all of the different bits of information it knows about you in combination, not in isolation. A loan company won't normally decline you just because of your job or just because you are a young adult[31]. The machine learning process considers all of your attributes in a holistic way to come up with an overall view of your creditworthiness. Everybody is treated individually, and the credit score you receive is probably based on a completely different combination of characteristics than everyone else. The way your credit score is calculated is unique to you[32].

Apply the same principle to politics and identifying the floating voters is what it's all about. Machine learning can predict who is likely to be a staunch Democrat or Republican voter. As a political activist I can discard these people from my campaigning and focus on the rest of the population, the floating voters, who have not yet made up their mind about who they will support, to try and convert them to my cause.

More widely, all of the applications of AI that we have discussed so far are based on similar machine learning principles. A hiring AI, used to identify good job candidates, predicts for each candidate, how likely it is that they will be a good fit for the role. The logic that this prediction is based on will have been derived from large samples of previous employee data. In particular, the features of each employee and how they have performed in their job. When assessing new candidates for a role, the candidate that the system predicts has the highest probability of fitting the bill "wins" and is offered the job.

In a similar vein, an AI-based object recognition system will have been trained using many thousands of images to identify the features that relate to different types of object. When the system is presented with a new image and asked what it sees, it creates predictions as to how likely it is that the pattern of pixels it's presented with is relates to each object. If the system predicts that there is a 5% chance that the image is a cat, a 15% chance that it's a cake and an 80% chance it's a potato, then the system will report that it sees a potato.

Advanced game playing AI technologies also apply this

approach. Each turn, a chess playing AI will assign each possible move a probability of leading to check mate, and select the move most likely to move it towards that state. These probabilities have been derived based on large numbers of historic games and how those games played out based on the moves that were made.

Technically, machine learning and artificial intelligence are not the same thing. Artificial intelligence is a much broader field of study than machine learning. Practically however, nearly every commercial AI application in the world today has some form of machine learning at its heart. Therefore, it's perhaps not surprising that many people use the terms "Artificial intelligence" and "Machine learning" interchangeably, even if it's not quite right to do so. In fact, this is something you will see that I'm guilty of throughout the rest of this book, but at least in doing so it should help to keep things simple.

Another way to think about this is that machine learning is one very powerful tool that can be used in the creation of AI powered apps, gadgets and robots, but it's not AI itself. There are other aspects to artificial intelligence that machine learning doesn't cover and there are also other tools available that can be used to develop AI applications such as **_Expert Systems_**.

The idea with an expert system is to create a database of knowledge gathered from the environment or provided by experts in a particular field. This database can then be interrogated to provide an answer to the question that has been raised.

Perhaps the most famous example of an expert system was a program called MYCIN, which was used to identify blood disorders[33]. To start, the user would answer questions about the patient's symptoms. MYCIN would then use a set of rules to infer things from those symptoms, via the information contained in its database. It would then draw a conclusion as to what blood disorder the patient was most likely to have.

A big benefit of expert systems is that you can, in theory, question thousands of experts and so create a database of knowledge that covers everything known about a given field of human endeavor. Typically, no one expert knows everything there is to know about their domain of expertise but in theory an expert system can. Therefore, it has the potential to outperform even the best individual expert.

There are some expert systems in use today, but they are only used in a very small number of applications and have tended not to be popular with the technology companies. There are a number of reasons for this, but a key one is the effort and cost required to build and maintain them. In particular, one has to spend a lot of time questioning experts and gathering data to create the database that the expert system requires.

However, most modern AI applications do owe something to the concepts underpinning expert systems. In particular, the use of experts to provide domain knowledge to enhance or over-ride a pure machine learning based approach. Many AI tools incorporate human domain knowledge, in the form of rule sets, to deliver something that is better than a machine learning or human-based approach could deliver on its own. In particular, rules derived from human expectation and/or expert opinion are used to over-ride machine learning based decisions to prevent AI applications making decisions that would be illegal or unethical, or in some cases, just downright stupid.

One example of an ethical rule that a human would define is where a marketing AI is targeting people on social media with adverts for alcoholic drinks such as rum. The AI is tasked with showing adverts to those people most likely to buy rum to encourage them to make a purchase. It might deem that some pregnant women would be good targets. Targeting pregnant women is perfectly legal and some of them may well want to drink rum. However, most people would agree that it is unethical to encourage pregnant women to drink alcohol because of the potential harm it could do to the unborn child.

There is also reputational risk for the company, in that if it becomes public knowledge that they've been targeting pregnant women, then that will not look very good for the company. Therefore, the marketing exec in charge of the AI would specify that the system needs to include rules such as: "Do not target pregnant women" to over-ride any decision that the marketing AI might make.

If we now take a legal perspective, a human defined rule that would also be required would be not to promote adult products (such as rum) to children. A common sense rule, would be not to target dead people or people in prison.

A highly publicized real-world example, that demonstrates the dangers of leaving everything to the algorithms, is the case of YouTube and its ad placement policy. YouTube uses a machine learning approach to decide what ads to show alongside which content. Many large organizations withdrew their advertising from YouTube when it was discovered that YouTube were placing some adverts alongside material from terrorists and other unsavory sources.

It was YouTube's machine learning algorithms that decided which ads to place where that had caused the problem. Consequently, YouTube had to undertake a major review of its ad placement process[34]. The result? After many months they eventually decided to revert to a manual vetting process to ensure that nothing slipped through the net. Machine learning on its own was not enough. Every video clip had to be reviewed and approved by a real person before it was included in YouTube's service which paired advertisers with popular content[35].

3. Algorithms, Models and more Algorithms

"We live in a society exquisitely dependent on science and technology, in which hardly anyone knows anything about science and technology." [36]

Algorithms. Algorithms. Algorithms. These are what drive all modern artificial intelligence applications in the world today. The media abounds with stories about how "Company X has developed an algorithm to do this" and "Company Y are using an AI algorithm to do that."

If you are not technically minded then algorithms can seem, well, a bit boring. However, if you want to really get to grips with AI tech and understand the basic principles of how it works, then grinding your way through this and the following two chapters should prove worthwhile. Get through these and it should all be plain sailing from there. However, if it all becomes a bit of a chore, then by all means skip ahead to Chapter 6.

Algorithms can be very complex and mathematical things but don't be put off by that. At its heart, an **algorithm** is just a set of actions or instructions, performed one after another, to complete a certain task or objective. You may think that algorithms only apply to computers and doing very clever things but that's not the case. Almost any everyday task or industrial process can be described using an algorithm.

Let's consider a relatively simple and mundane task such as making a nice hot cup of tea. The following is an algorithm that describes the set of actions that I undertake when I make myself a cup. As you can see, it's not particularly complex or difficult to understand.

---

1. Put a teabag in a cup.
2. Boil some water.
3. Fill the cup with the hot water.
4. Wait a few minutes for the tea to brew.
5. Take the teabag out.
6. Add milk and sugar to taste.
7. The tea is now ready.
8. END of algorithm (now drink the tea!)

This algorithm captures the way that I go about making a cup of tea. You may do things somewhat differently. Maybe you're a bit of a connoisseur and use loose leaf and a fancy teapot; or maybe you're a bit of a slob who just leaves their teabag in the cup while they drink it – I know lots of people who do that!

There is often more than one way to climb a mountain, and in practice, there is rarely just a single algorithm for achieving a given task. Many different variants are possible, each with their own strengths and weaknesses. Some algorithms will be more efficient, others faster, other more flexible. They all approach the problem in a slightly different way.

When it comes to computer algorithms, the primary difference is that the language used isn't plain English like we've used in the tea making example. Instead, a more stylized and precise type of language is used that a computer can understand – what is commonly referred to as computer code.

Just like human languages, there are lots of different computer languages. Maybe you've heard of some of them (or maybe not, if you're not a nerd or geek):

- Python
- Visual Basic
- JavaScript
- SQL
- Rust
- C++
- PHP

These are all examples of different types of computer language that allow algorithms to be expressed in different ways. Just as it's easier to express certain ideas and concepts in some human languages than others, so each computer language is designed to be good at certain things.

If you want something that is going to be really fast and efficient at performing an algorithm, then C++ is often the language of choice. The weakness of C++ is that it has quite a steep learning curve. Alternatively, for something that is easy to understand and use, and can be learnt quickly, then languages like Python or Visual Basic are probably better. However, the downside is that these languages can be a bit slow if you have very complex algorithms that need a lot of computer power to complete.

Despite their differences, the one common feature of all these computer languages is that they need to be very precisely written in order that the computer executes exactly what the computer programmer intends without any confusion or ambiguity.

If you think about my tea making algorithm, its full of ambiguity. There are loads of assumptions and simplifications that a person makes without thinking, based on their common sense, which a computer doesn't have. Human common sense allows us to execute the algorithm successfully, without having to be explicit about every little detail.

If I were to rewrite the tea making algorithm in a form that is something a bit more like computer code, then it would be more along the lines of this:

1. Get 1 cup, with a capacity of between 0.30 and 0.40 litres.
2. Put 1 teabag in the cup.
3. Get kettle.
4. Empty kettle (of any old water).
5. Fill kettle with 0.25 litres of cold water from kitchen tap.
6. Turn kettle on.
7. Heat kettle until water boils.
8. Turn kettle off.
9. Pour the water from the kettle into the cup.
10. Wait for 3 minutes.
11. Remove the teabag from the cup.
12. IF milk required THEN add 25 ml of milk (0.025 litres).
13. IF sugar required THEN add 15 grams of sugar.
14. END of Algorithm.

Now, this isn't proper computer code, but hopefully it demonstrates my point. Everything needs to be clear and exact. It's pretty obvious to you or me that you just fill the cup until nearly full, leaving space for milk if required. That's all we need to be told for us to complete the task successfully. Our innate common sense provides the rest. For a computer, unless you specify exact sizes and quantities, you are likely to get into trouble. If instruction 7 had said "Heat water to 100 degrees centigrade" instead of "Until water boils" then the algorithm would never finish unless it was working at sea level or below. For anyone above sea level, the kettle would boil dry and the tea would never get made. Why? Because the water would never reach 100C. The boiling point of water drops significantly once you are just a few hundred meters above sea level[37].

OK. Now that we know what an algorithm is, let's get back to artificial intelligence and machine learning. In the world of AI/machine learning, there are several different types of algorithms that are applied in a number of different ways, at different times. Consider Figure 1.

Figure 1. Algorithms used in an AI application.

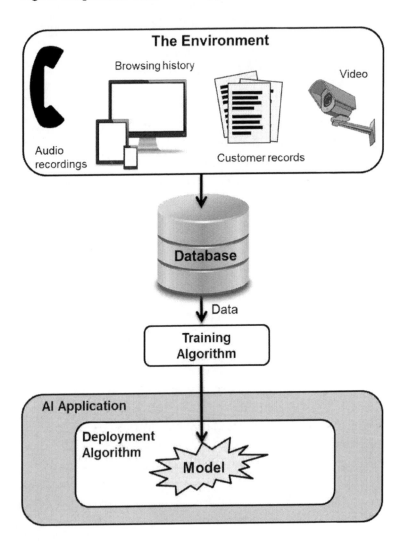

Before any algorithms are applied, data about the environment, **training data**, needs to be collected together in one place. In Figure 1, this is represented by the **database**. The database containing the training data may be located on your desktop PC or laptop, a company server or "In the cloud" somewhere i.e. a third-party organization that offers data storage services without you needing to know exactly where the data is being held.

Once training data has been gathered and stored in the database, then we can start to apply the algorithms. As shown in Figure 1, there are three main types of algorithm involved in the development and use of a typical AI application.

1. **Training Algorithms.** The instructions carried out as part of a training algorithm use the training data to learn about the relationships that exist in the environment. Most commonly, we are interested in how one specific environmental factor is influenced by all of the other factors. The **model** produced by the training algorithm captures all of the relevant patterns and relationships that the algorithm has found.

2. **Model Algorithms.** The model produced by the training algorithm, is itself, a type of algorithm. The logic contained within the model provides a complete record of all the relationships that the training algorithm has discovered about the environment.

3. **Deployment Algorithms.** The final type of algorithm provides the operational framework within which the model is placed to enable it to be put to use; i.e. for actions to be taken based on new data that is presented to the model about the environment.

With an AI-based app, it's all about the model and how you use it. The training algorithm is vital, and is usually the most complex and computer intensive part of the process, but it's just a means to an end. Once we have a model, then we don't need the training algorithm any more (well, at least not until we want to develop another model).

Right ho, that's all very well. Models, and how you use them, are where it's at when it comes to AI, but what exactly is a model? What does a model look like and how does it produce predictions or other outputs?

The Compact Oxford English Dictionary provides seven definitions of what a model is. However, we are not talking about going down the catwalk or a 1 in 25 scale model of a Falcon Heavy rocket. The definition of a model that we are interested in is as follows:

> "A simplified mathematical description of a system or process, used to assist calculations and predictions."

In other words, a model is a mathematical equation (a function), or set of equations, that captures how the environment changes or reacts under a given set of circumstances[38]. If you provide the model with a set of inputs, then these will be processed by the model to produce one or more outputs. The outputs represent a view of what the inputs mean in relation to some aim or objective that the system has.

For want of a better analogy, and without meaning to go all "Mad scientist" on you, the model produced by the training algorithm is the "Brain" that provides the "Intelligence" in an AI application. Figure 2 provides an illustration of how this happens.

When the deployment algorithm presents data to the model (The inputs in Figure 2) the model processes it and produces outputs. The outputs are then picked up by the deployment algorithm and different things then happen based upon what those outputs are.

Figure 2. Model operation.

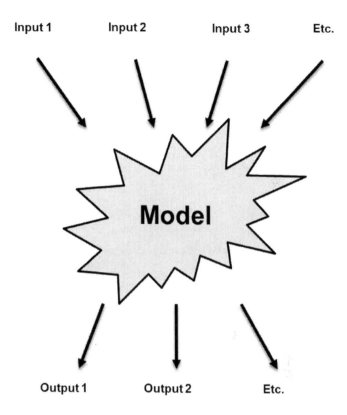

Many machine learning models produce just a single output. Some produce several outputs and a few tens of thousands. However, regardless of the number of outputs a model produces, each output is normally a number, or score, that can be one of two types.

- **Probabilities.** Each of the model scores is a prediction that estimates how likely a certain outcome is to occur.

- **Magnitudes.** The scores are predictions about the size or magnitude of something. For example, how long a car journey will take or how much a customer will spend.

Predictive models are a core component of nearly all AI applications, but there are other elements required to deliver a fully functioning AI app. The model needs to be deployed into an environment where it can interact with that environment so that things happen when decisions get made. In particular:

- You need to decide what decisions to take, based on the scores that are produced.

- Once a decision has been made, it needs to be acted upon. Decisions and actions are not the same things.

Going back to our brain analogy, you have to put the brain into a body and connect it to the sensory organs and muscles via the central nervous system. When the brain decides to raise a hand, blink or say something, the decision made by the brain needs to be communicated to the rest of the body where the action occurs. The brain on its own, without those connections to the outside world, is a pretty useless thing. We'll talk more about model deployment later. For now, let's focus a bit more on the model element.

To further understand what a model is, as with most things, it helps to put things into context with a couple of concrete real-world examples. However, rather than jumping into the real cutting-edge stuff (don't run before you can walk), let's start with some pretty mundane, old school and very mainstream examples, that many people may have had some experience of from a customer perspective.

For our first example of AI in practice, consider the task of predicting who is most likely to develop certain medical conditions, such as diabetes, in the future. The reason why you want to predict this is so that the people most likely to develop diabetes can be targeted for help, to take preventative action now, to reduce their risk of developing the disease before it occurs. Prevention is better than cure for all sorts of reasons, not least: patient well-being, the cost of treating diabetes and increased productivity at work.

The environmental data that the training algorithm is going to use in this example, is a database containing details of thousands of peoples' medical records. The aim/objective is to establish the relationship between developing diabetes in the future and all of the medical information that we have about patients today.

We obviously don't know the future yet. Therefore, the starting point is historic medical records from a few years ago for people who *didn't* have the condition (there's no point predicting if someone is going to develop a disease if they already have it!) This is then augmented with information about what happened to their health over the next few years, and in particular, each patient record is flagged to indicate which patients developed diabetes and which did not.

Figure 3. A decision tree model for predicting diabetes.

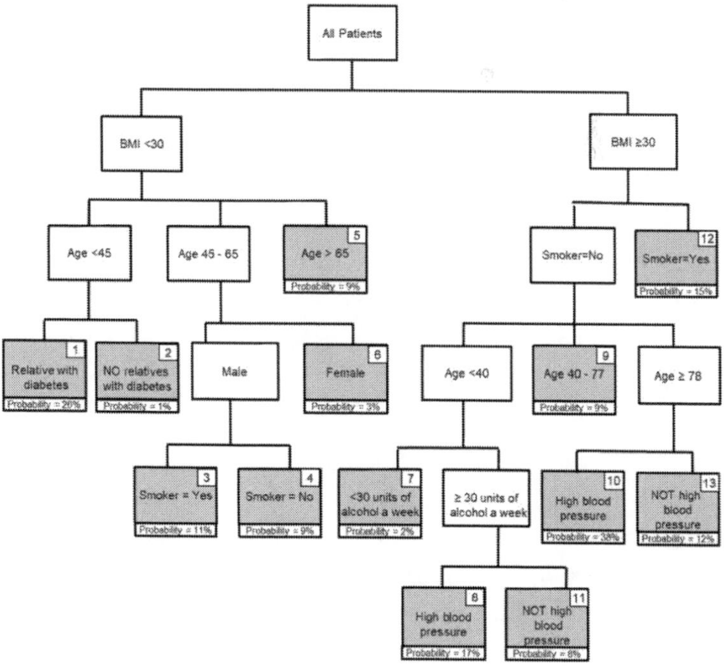

The training algorithm then trawls through all of the patient data, looking for all the things in the patient records that are related (correlated) with developing diabetes. The things that the training algorithm finds are captured by the (predictive) model at the end of the process. The model records the relationships between the patients' details in the past and if they subsequently develop diabetes a few years later, as shown in Figure 3.

This type of model is called a ***Decision Tree***. The algorithm used to derive the decision tree may have contained a lot of complex math, but the way the decision tree model works is pretty simple, as follows:

1. Start with the first node at the top of the tree (All Patients).

2. At the branch point, choose the branch that leads to the node that fits the patient's data. For example, choose the left branch if the patient's Body Mass Index (BMI)[39] is less than 30.

3. Repeat the process for each branch, until you get to a final "Leaf node" (I think you get why they call it a tree) and can go no further (One of the numbered shaded boxes).

4. Read off the probability associated with that node. That's our prediction as to how likely someone is to get diabetes.

The decision tree has 13 leaf nodes. A patient with a given set of characteristics will end up at one, and only one, of these nodes. The leaf node into which a patient falls drives a single output: The probability of developing diabetes. If we take someone with the following characteristics:

- Body Mass Index <30.
- Age 50.
- Male.
- Drinks <30 units of alcohol a week.
- Does not have high blood pressure.
- No relatives with diabetes.
- Smoker.

Then this person will end up at leaf node 3. Therefore, they are deemed to have an 11% chance of developing diabetes.

How are the probabilities associated with each leaf node determined? It's pretty straightforward. You just count up how many cases with/without diabetes ended up at each leaf node, and do some simple arithmetic.

Let's say that 5,000 patients end up at leaf node 1, and of these, 1,300 went on to develop diabetes. That gives us a probability of 0.26 (1,300 / 5,000) or 26% if you like to think about things that way.

The key assumption here is that outcomes for people with a given set of features will be the same as it was for patients historically. The patterns seen in the past will be repeated in the future. Therefore, knowing how things played out previously can tell you a lot about how things will go this time around. That's a pretty big assumption to make, but the evidence is that it tends to hold true in all sorts of situations, with the odd exception.

Easy! Anyone can understand and use a decision tree model like this. Feel free to take a copy to use at your next dinner party[40]. However, the real magic happens when we deploy the decision tree model in the real world such as a GP's surgery. The GP can present the model with up-to-date medical records from all of their thousands of patients *without* diabetes today. The model then generates, for each patient, a probability of them developing diabetes in the future based on the leaf node of the decision tree into which they fall.

This is a very powerful thing to be able to do, because it means that in a world where health care resources are limited, people who the model predicts have the highest chance of developing diabetes can be targeted for individual help. So, the GP might, for example, decide that they have enough free time to be able to invite all of his patients who fall into leaf nodes 1 and 10 to attend a lifestyle review. This is because these people have the highest chance of developing the disease. That's not to say that if you end up in node 1 or 10 you will definitely develop diabetes, but that your chance of doing so is very much higher than everyone else. Hence, the benefits of targeting them for treatment will be much greater than targeting other groups within the population.

Because it's all on a computer, the model can be applied to thousands of patients' records in seconds, which means that the GP can run this process as often as they want. Every patient could be reassessed every day if necessary, and if their chance of getting diabetes becomes very high, then they can be added to contact list and invited to visit the surgery where appropriate advice can be given.

Things start to become even more interesting when we start to incorporate information into the process that a doctor would not traditionally have had access too. For example, biometric data from smart watches or other similar devices, or data about what food people are buying at the supermarket each week. This type of rapidly changing real-time data often provides huge insights into people's behaviour in all sorts of different ways, including their health.

That's a healthcare example of machine learning. Let's now go back to the world of credit granting and credit scoring, where machine learning has been used for decades. Imagine that I work for a financial services company that provides personal loans. My boss wants me to rebuild their "AI-driven underwriting application." Customers can download the app on their phone, which then automatically decides which customers applying for a loan should be given one.

The company has been issuing loans for several years. Consequently, it has a database containing information about all of the thousands loans it granted in the past. This database of historic loans holds two key types of information about previous customers that are going to be very important for developing the underwriting app.

1. **Application data.** This is information that was known about people when they applied for their loan. This will include stuff they provided themselves, such as their age, income and employment status, but also lots of other information that the loan company had access to from other sources, such as credit history from a credit reference agency or social media data.

2. **Repayment data.** This is information about loan repayments. In particular, did the customer repay their debt or did they default, resulting in the loan written-off as a bad debt?

The aim/objective in this case is to use information contained in the application data to infer things about individuals' subsequent repayment behaviour. To put it another way, can we create a model that uses the application data that was supplied when someone applied for a loan as its inputs, and produce an output that predicts their chance of default if we did decide to lend them the money?

To produce a model of the underwriting environment, I would then apply a suitable training algorithm to the database of historic loans. This is having first set up the training algorithm such that it knows that my objective is to predict loan repayment behaviour, using the application data. The algorithm would then trawl through all the thousands of examples of historic loans in the database to establish which items predict loan repayment behaviour and in what way.

The end result is a **_Scorecard_** type model, as shown in Figure 4.

Figure 4. A scorecard model.

Starting score	700
1. Annual income	
<= $25,000	-21
$25,001 - $30,000	-14
$30,001 - $40,000	-9
$40,001 - $60,000	0
$60,001 - $85,000	5
$85,001 - $120,000	9
$120,001 - $160,000	25
> $160,000	32
2. Employment status	
Unemployed	-42
Full-time or retired	28
Part-time	7
Homemaker	28
Student	-8
3. Time in current employment	
Not in full or part-time employment	-14
<1 year	-25
1 - 2 years	-10
3 - 8 years	0
> 8 years	31
4. Eye color	
Blue	4
Green	0
Brown	-3
Other	0
5. Residential status	
Home owner	26
Renting	12
Living with parent	0
6. Number of credit cards	
0	-17
1 - 2	0
3 - 4	-5
5 - 7	-11
8+	-24
7. In arrears with any existing credit agreements?	
Yes	-38
No	0
8. Bankrupt?	
Yes	-62
No	0

In coming up with the scorecard in Figure 4, the training algorithm has examined all of the data it was presented with. This may have included hundreds or thousands of data items about previous customers. However, the algorithm has concluded that the 8 items in the scorecard are the only important ones for assessing credit risk in this example.

This is not an unusual situation. Very often, only a very tiny fraction of the available data proves useful for any given task but you don't usually know which data items are important in advance. Therefore, you have to analyze pretty much all of it when creating a model if you want to get the model to be as accurate as possible. Finding which bits of data are useful, and which are not, is part of what the training algorithm does[41]. In this example, there may have been information about marital status, educational achievement, tweets people had posted, where people live and so on but these items where not found to be important when predicting loan repayment behaviour.

If you want to work out your credit score using the scorecard model in Figure 4 then it's pretty straightforward. All you need to be able to do is to add and subtract the values that apply by using the following algorithm:

1. Give yourself the starting score of 700.
2. For each piece of information in the scorecard model, find the score that applies to you.
3. Add up all the scores that apply.
4. Hey presto, that's your credit score.
5. END of Algorithm

For someone with the following attributes:

1. Has an annual income of $38,000.

2. Is in full time employment.

3. Has been in their current employment for 2 years.

4. Has green eyes.

5. Is a home owner.

6. Has two credit cards.

7. Is not in arrears with any existing credit agreements.

8. Is not bankrupt.

They would start with a score of 700. You then subtract 9 points due to their income, add 28 points for their employment status and so on to get a final score of 735.

The higher the score, the more creditworthy you are; i.e. the more likely you are to repay any money you borrow and the less likely you are to default. Conversely, the lower the credit score the worse credit risk you are.

The strength of the relationships are reflected in the magnitude of the scores. If an attribute has a larger positive or negative score than another attribute, then that indicates that that attribute is more predictive of default and therefore contributes more to the final score that someone receives.

Many of the things in the scorecard align with common sense. If someone hasn't been in their job very long or has recently been declared bankrupt then it's reasonable to assume that they aren't as financially stable as people on high incomes or who have been in their job a long time. Therefore, they should get lower scores to indicate that they are more of a risk.

OK, so why can't I just get an experienced underwriter to come up with a scorecard instead, based on their expert opinion? Why do I need an algorithm? Well you can, and the scorecard that they come up with would probably be reasonably good at assigning credit scores to people that reflect their creditworthiness[42]. In fact, when credit scoring was first introduced, this was one of the arguments put forward against its use. However, there are two things that training algorithms do which tends to result in scorecards derived using algorithms being better than those derived by human experts.

First, the scores they assign to each attribute are optimal. By optimal, I mean that they are as good as they can be, based on the data used to train them. Giving someone -42 for being unemployed is better than giving them -41 or -43. i.e. -42 best reflects the risk from being unemployed. Second, and perhaps most importantly, the algorithm can find completely unexpected relationships in the data that can be used to predict a person's creditworthiness.

These two elements, the process of identifying what data items are important and what scores each attribute receives, are the clever bit – that's where all the fancy math occurs.

In this example, the unexpected thing that the algorithm has included in the scorecard is eye color. It has determined that people with blue eyes are a slightly better credit risk than people with other eye colors. This brings us on to a very important danger that exists when machine learning training algorithms are applied. All they do is find relationships in data. They don't have any insight into the meaning of that data, why those relationships exist or the context within which the data was collected.

Without a degree of human oversight, there is a very real risk of unethical and/or illegal biases being introduced. Why? because if the training data reflects biases that exist in wider society then these will be mirrored by the training algorithm. Eye color may be predictive of creditworthiness, but eye color is also highly correlated with ethnicity, and there have historically been a lot of prejudicial bias against certain ethnicities which may be reflected in the data. It's therefore ethically questionable to include a variable such as eye color in a predictive model used for this type of commercial purpose. Consequently, the data scientist in charge of building the model (i.e. me) should ensure that eye color is excluded from the data used by the training algorithm and hence does not feature in the final model that results.

Some might argue that by removing eye color from the scorecard it will not be as predictive as it could be, and they would be right, but that's the price one has to pay to ensure that the model is fair, and in many regions of the world, legal too. We'll talk more about this issue of bias and the ethics of AI later.

In terms of traditional credit scoring, that's pretty much it. The scorecard in Figure 4 is a somewhat simplified example that uses only 8 pieces of information, but organizations all over the world have been using similar types of scorecard models in their underwriting for decades. Many contain no more than 2 or 3 dozen items of data at most.

A credit score is a measure of creditworthiness. High score good, low score bad. But how does an organization use the scorecard? How does it decide what a good or bad credit score is, and most importantly, how high do you have to score to get a loan?

(WARNING – mathsy bit coming up). Let's now imagine that I calculate the credit score for all the thousands of customers that previously had a loan. I can then use the scores in conjunction with the information about loan repayment behavior to generate a customer score profile as shown in Figure 5.

Figure 5. Interpreting model scores.

Score Range	Percent of customers who repaid their loans	Percent of customers who defaulted
0 - 550	55%	45%
551 - 600	77%	23%
601 - 650	83%	17%
651 - 700	88%	12%
701 - 750	91%	9%
751 - 800	96%	4%
801 or more	99%	1%

In Figure 5, the leftmost column shows the score range. The rightmost column the proportion of people who subsequently defaulted on their loan. For our example, for the customer who scored 735, they fall into the 701 – 750 category. In that category, 9% of customers with those scores defaulted on their loans in the past. From that, we estimate that our customer, applying for a new loan today, also has a 9% chance of defaulting if we grant them the loan.

Now for the tricky part. Let's say, that having examined the profit and loss of previous loans, the company has concluded that:

- For each loan that is repaid, it makes an average of $500.

- For each case of loan default, it loses an average of $4,500.

That means, that to make a profit, they have to grant at least 9 goods loans that repay, for every one that defaults; i.e. 9 lots of $500 to cover the $4,500 loss. Anything better than that is profit. To put it another way, they can afford an average of 1 loan default for each 10 loans granted, which gives us a break even default rate of 100% * 1/10 = 10%

To make money, the average default rate needs to be less than 10%. Referring to Figure 5, we can see that the default rate is less than 10% if customers are scoring 701 or more. Based on this information, the **decision rule** or **threshold** to apply is very simple:

- Grant loans to applicants scoring 701 or more.

- Decline loans to applicants scoring less than 701.

That's a very simple example, but the idea of having a threshold, or cut-off, is fundamental to all sorts of predictive models used in practice and in AI in general. If the model output (the score in this example) is above a certain value do one thing, if it's below a certain value do something else.

In retail credit, different loan companies use different models and operate at different levels of profitability. Therefore, each company will have its own view as to what credit score it deems acceptable and use that within the decision rules that they use to decide who to lend to. Some companies won't lend to anyone where the default rate is expected to be anything more than 1%, which covers all those low rate mortgages and credit cards. Move into the payday loan market, and a 50+% default rate is acceptable to some lenders.

In terms of how well an automated loan granting system, based on a scorecard, performs then the general findings are that they can outperform trained human underwriters by 20-30%[43] To put it another way, by replacing human underwriters with the scorecard, a loan company can expect to see a reduction in bad debts of 20-30% and every dollar of reduced bad debt is another dollar on the bottom line. Once organizations understood the potential benefits, credit scoring was a no brainer.

OK. Hopefully you get the idea. In both the diabetes and credit scoring example, the same type of logic applies. If the model output (the probability or score) is above a certain value take one type of action, if its below take another. As we'll see in the next chapter, this type of decision-making, based on the scores produced by a predictive model, underpin almost all of the Artificial Intelligence applications in use today.

A key feature in the development of these types of models is that the process can incorporate adaptation and continuous improvement. It can learn and change over time. As the company grants more loans or the doctor sees more patients, so the amount of information about historic cases increases. If the training algorithm is applied again, when more data is available, then it will produce a new, different, more accurate decision tree or scorecard.

In theory, there is no reason why these models can't be updated all the time, automatically in real-time, each time a new piece of data becomes available[44]. The more data a training algorithm has access to, the more predictive the resulting models become. As a general rule, a very simple training algorithm provided with a lot of good quality data will produce a much more accurate model than a very sophisticated training algorithm using less data.

Scorecards, decision trees and more advanced variants of these types of models, such as **ensembles** and **random forests,** are great. They are used all over the place in all sorts of processes, in many different organization from tax authorities, to pharmaceutical companies to law enforcement. I've even seen a scorecard used in A&E to support amputation decisions. Score high enough and they cut it off!

These types of model are often easy to understand and explain, and new, updated versions, can be generated as often as you want. For these reasons they remain the most popular types of model in general use[45]. However, they do have their weaknesses and are better suited to some tasks than others. In particular, scorecard and decision tree type models don't tend to perform particularly well and/or aren't as good at decision-making as people when they are applied to tasks where:

1. The environment is complex; i.e. There is a very large amount of data available that contains very complex, nuanced and inter-dependent relationships within the data.

2. The data available to the training algorithm is predominantly unstructured; that is, data that can't easily be formatted into a nice row/column format like a spreadsheet. Examples of unstructured data are pictures, news stories and sounds.

3. They are used to try and predict multiple complex outcomes. For both the diabetes and credit scoring models, there was only one output predicting a single thing (getting diabetes or repaying a loan respectively).

In the past, when not much data was available, and what data did exist tended to be in a neat, structured, tabular format (like one finds in a spreadsheet such as Microsoft Excel), this wasn't so much of a problem. However, in todays' world we are awash with data, much of it messy and unstructured such as text and video. What's been seen is that as the amount and variety of data has grown, even the best algorithms for producing scorecard and decision tree type models struggle to deliver good solutions. Some examples of complex tasks that it has proven difficult to get good results for using scorecard and decision tree type models (and quite a few other types of model as well) include:

- Playing games. This includes traditional games such as Poker, Chess and Go, as well as modern computer games like StarCraft and Quake.

- Identifying and describing items contained in pictures.

- Speech recognition and language translation.

- Creating advertising copy; i.e. writing text to promote a product or service when you look it up online.

- Facial recognition.

- Answering diverse customer queries on websites (chatbots).

To be able to equal or exceed human abilities in these fields, something more flexible, more adaptive and just downright more powerful is required. Something biologically inspired…

4. The AI Explosion. Neural Networks and Deep Learning

"Thou shalt not make a machine in the likeness of a human mind."[46]

Billions upon billions of neurons with trillions of connections between them. That's what drives our intelligence and our ability to learn and adapt when we encounter something new, that we have not experienced before. It's the combination of neurons and connections that provides our ability to understand our environment and how the world works. This in turn drives our ability to make decisions and guides how we interact with each other and the wider world.

As we age, develop and grow, we are exposed to new and different situations. To be able to deal with these new situations, the neurons and connections in our brains change in response to the stimuli they receive. They revise their learning to incorporate the new information that is constantly arriving via our senses as we move through our lives. In this way, the decisions about what to do and how to behave reflect the ever changing world that we inhabit.

It was way back in the 1950s that computer scientists first came up with the concept of an artificial neuron[47] that simulated, in a very simplistic way, the behaviour of a natural biological neuron. However, it wasn't until the middle of the 1980s that the idea of using several artificial neurons linked together, a ***neural network***[48], was proposed as a way of representing and solving complex problems[49]. In the years that followed the popularity of neural network models slowly continued to grow but they remained a fairly niche area of scientific research until around 2010. It has only been in the last few years that the necessary computing capabilities have become readily available to allow their full potential to be realized and for them to be incorporated into everyday consumer devices and business processes. Today, advanced forms of neural networks are at the cutting edge of artificial intelligence and machine learning research.

The first thing to say about neural network models is that they are often touted as being immensely complex "Brain like" things, but they can be readily understood if you are willing to spend a little time and effort to study them. It's certainly true that a typical neural network is a lot more complex than the scorecards and decision trees that were introduced in the previous chapter, but the underlying principles are not that much different at the end of the day – there is just more of it.

To show how a neural network works, let's return once more to the credit scoring example that was introduced in the last chapter. If you remember, this was where we used a scorecard type model (Figure 4) to rate people based on their characteristics such as their income and employment status. High scoring "Good" people, who were deemed to be very unlikely default on their debt, would be granted loans. However, those with lower scores were judged to be too high risk, and would therefore, have their loan applications declined. This time however, we'll generate peoples' credit scores using a neural network model rather than a scorecard model.

We are going to start with the building block of a neural network – the **neuron**. However, before we do that, let me put any fears you may have to rest. In the world of artificial intelligence, a neuron is not a living thing. It's not some throbbing blob of sticky brain matter, grown in a clandestine black-ops lab by a team of evil scientists. Rather, a neuron is a very simple and simplistic representation of how natural neurons behave, created using equations and implemented as computer code. It's just another type of algorithm. Figure 6 provides an illustration of how an artificial neuron works, based on the credit scoring example introduced in the previous chapter.

Figure 6. An artificial neuron.

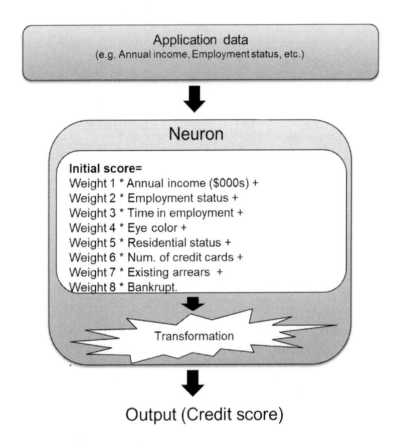

The way the neuron in Figure 6 works is as follows:

1. The data gathered about each loan application provides the inputs to the neuron. This is just like in the scorecard and decision tree models.

2. Each input is multiplied by a weight (a positive or negative number). For non-numeric data, such as employment status, numeric indicators or flags are used to represent each option. 1 = employed, 2 = retired so on[50]. If Weight2 was say, 0.75, then if the applicant was retired, that would contribute 1.5 to the initial score (0.75 * 2).

3. The inputs, multiplied by their weights, are added together to get an initial score.

4. The initial score is subject to a transformation to force it to be in a certain range[51], often between 0 and 1. This is so that when several neurons are combined together to produce a neural network, all the neurons produce values in the same range; i.e. between 0 and 1.

5. The transformed version of the initial score is the output produced by the neuron; i.e. the credit score.

A neuron in the context of machine learning isn't anything mysterious or complex. Even the mathematics involved is not too hard, it's just a couple of simple formulas:

1. The first multiplies each input variable by a weight and adds up the results.

2. The second formula is the transformation (what's called an ***activation function***[52] in data science speak) which modifies this value to lie within a fixed range.

There is nothing more complex to an artificial neuron than that! The clever bit is determining what the weights should be, which we will cover shortly.

On its own, a single neuron is not really any different from the scorecard type model we introduced in the last chapter. In fact, it can be demonstrated that the outputs generated by a single neuron and a scorecard type model are equivalent[53]. The only material difference between the two models is the transformation undertaken by the neuron to force the score to lie in a fixed range.

To produce a neural network model, a number of neurons are connected together. Figure 7 provides an example of how this occurs.

Figure 7. A neural network model.

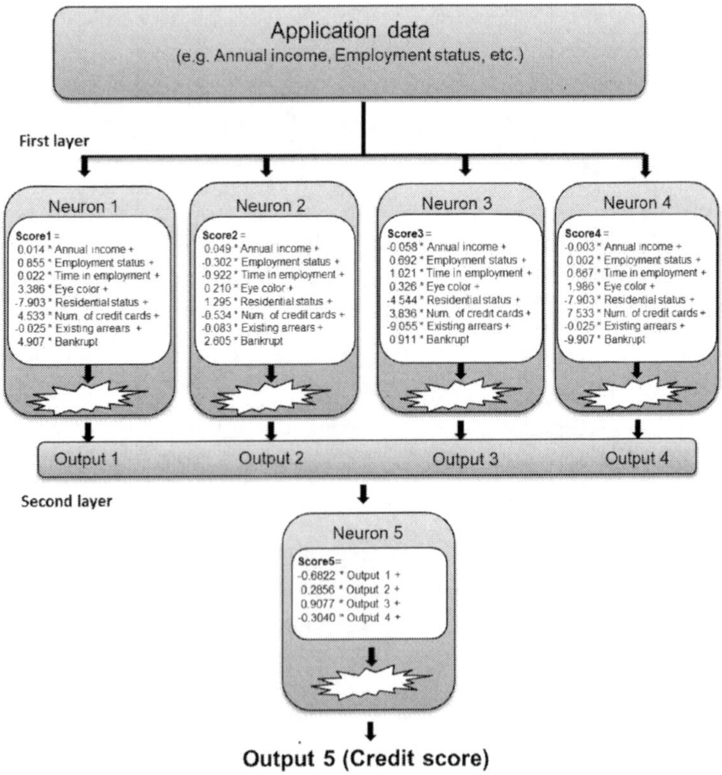

Output 5 (Credit score)

In Figure 7, the Credit Score (Output 5) from the network is calculated as follows:

1. The application data is supplied separately to each of the four neurons in the first layer.

2. Each neuron has its own weights. The weights, when combined with the application data, create scores (Score1, Score2, etc.)

3. The scores are transformed to produce four outputs.

4. The four outputs provide the inputs to the single neuron in the second layer (Neuron 5).

5. Neuron 5 combines the inputs with a further set of weights to create Score5.

6. Score5 is transformed to create the final credit score.

Some key features of the neural network in Figure 7 are:

- The inputs to each neuron in a given layer are the same, but each neuron produces a different output (score). This is because the weights in each neuron are different.

- The outputs from the first set of neurons (the first layer) provide the inputs to the second set of neurons (the second layer).

- There are four neurons in the first layer in this example. In practice, the number of neurons can vary considerably, and much trial and error is required to find the best number to use for any given type of AI application.

- There are quite a lot of weights. Figure 7 is a very simple neural network model, but it's clearly a lot more complex than the scorecard and decision tree models.

- The raw calculations used in the neural network are easy to understand but the weights don't make much sense. It's not easy to determine precisely what's important in determining the final score and what's not.

If we take a single loan application, then the credit score produced by the scorecard and neural network models won't be exactly the same. This is because the algorithms that create each of the models consider the data differently. However, the way in which the credit scores are used is identical regardless of what type of model is employed. All the lender is interested in is having the most accurate credit score possible, and then deciding if it's worth granting the loan. Arguably, it doesn't really matter what type of model is being used behind the scenes if the credit score it generates is a good one. Likewise, if we built a neural network to predict diabetes, then the probabilities generated by the network model would be used in just the same way as the probabilities generated by the decision tree. Different means, same ends.

The collection of weights in the network represent the patterns that the *training algorithm* has identified. If you want a biological analogy, the weights can be thought of as a memory of the learning that has taken place, which can be recalled whenever you want to make another credit scoring decision.

How does the training algorithm determine what the weights should be? There are many different algorithms that have been developed to find the weights in a neural network, but they all tend to adopt the following principles:

1. Assign each weight a random or zero value.

2. Calculate the scores generated by the network for all of the cases in the training data.

3. Assess how accurate the final score is; e.g. for the credit scoring model, how well the model assigns high scores to "Good" payers who repay their loans and low scores to "Bad" payers who default.

4. Adjust the weights so as to improve the accuracy of the model. For the credit scoring example, so that more of the good payers get higher scores and more of the bad payers get lower scores.

5. Repeat steps 1-4 until no further significant improvement in model accuracy is observed, or a certain amount of time has elapsed.

The clever bit, that involves all of the complex math, is in Step 4, how one adjusts the weights – a process referred to as **training**. The simplest training approach is to just randomly try different values and see what works best. However, this is very inefficient and is unlikely to yield a good solution in realistic time. Even the most powerful computer could run for years and still not find a good model using this approach.

All practical neural network training algorithms are cleverer than this. They adopt a variety of different weight adjustment strategies based on the differences in performance between each iteration of the algorithm. They will make big adjustments to the weights at the start of the process when it's easy to obtain large improvements in predictive performance, and then gradually make smaller changes as performance improves to home in on the optimal solution.

Ideally, the training algorithm would keep going until the model was perfect; that is, 100% accurate. For the credit scoring case, this would mean that all of the good payers would get the very highest scores and all of the defaulters would get the very lowest scores. In practice however, no model is perfect. Therefore, the training algorithm will typically end when further changes to the weights deliver no further significant improvements in the accuracy of the model.

As a short aside, there is an important lesson here. Predictive models are often better than human experts at predicting things such as creditworthiness or if you are going to get diabetes, but they aren't perfect. It seems to be in people's nature to pick up on cases where the models get it wrong and use this as an argument against their use, but at the same time ignoring the fact that the models are getting things wrong less often than humans. Perhaps the most significant evidence of this type of thinking is when it comes to self-driving cars. There are thousands of accidents on our roads caused by human error every day and almost no one blinks an eye, but a single accident caused by a self-driving car making a mistake is enough to send the media into a frenzy! It seems that we expect higher standards from AI than we do ourselves, which may be no bad thing.

OK - back to the main theme. One reason why neural network models are so popular is their ability to detect subtle patterns in data that other simpler methods, such as scorecards and decision trees, may not be able to detect. This means that they can potentially produce more accurate predictions, particularly where there is a lot of very complex data to consider. This sensitivity to subtle patterns in the data is achieved by having the two layers of neurons, with the scores from the first layer providing the inputs to the second layer.

The main drawback of neural networks is that the scores that they generate are not intuitive. Yes, one can see what the various weights in the model are, and understand the overall way that the credit score is calculated. However, if I were to ask you which of the data items (the inputs) in Figure 7 contributes the most to the final score, then this is much less obvious than for the scorecard model (Figure 4). This is potentially a problem if there is a business or legal requirement to explain how the final model score was arrived at.

Deep learning represents the latest evolution of neural network type models. The neural network in Figure 7 has two layers of neurons and this structure is used very successfully in many traditional neural network applications. However, there is no reason why there can't be many more layers. Have a look at the network in Figure 8.

The network in Figure 8 has the same initial inputs as the network in Figure 7. There is also a single output neuron which delivers the final model score (Neuron 13). From a user perspective, the two models take exactly the same input data and deliver the same type of output. However, the network in Figure 8 has 4 layers of neurons compared to just 2 layers for the network in Figure 7.

In theory, there is no reason why the network could not be extended even further. There could be 5, 6, 7, …, 100+ layers if desired, with different numbers of neurons in each layer. What you tend to find is that as more layers are added, so the ability of the network to identify complex and/or subtle patterns increases. The more layers the deeper the network. Generally speaking, anything with more than 2 or 3 layers can be classified as a deep network, but there is no single accepted definition.

The neural networks in Figures 7 and 8 have a single neuron in the final layer which generates a single output – a credit score. Another big strength of neural networks is that they can easily be structured to have many outputs, not just one. This is very important for tasks where there can be thousands of options to choose from, each requiring a separate output.

Figure 8. A deep neural network.

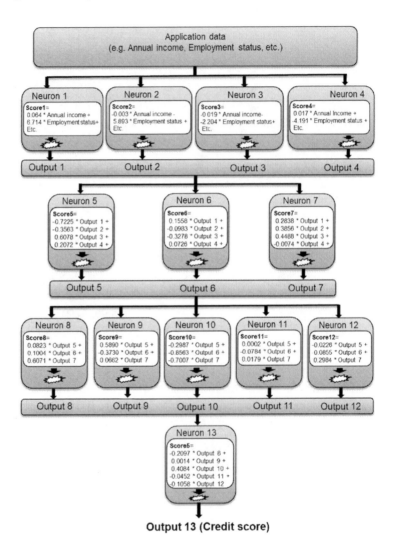

Output 13 (Credit score)

Let's go back to our object recognition task and think about how a neural network might be used to identify what object is in a picture. As before, we need to gather lots of information from the environment for the training algorithm to use. In this case, let's assume that the training data we have available to train the network is several thousand pictures of cats, cakes and potatoes, formatted as follows:

- **Input data.** Each picture is represented as a set of pixels. Each pixel has four pieces of data associated with it. A red, blue and yellow component to indicate the colour of the pixel, plus a value to represent the intensity (brightness) of the pixel.

- **Category data.** Each image is labelled to indicate if it contains a picture of a cat, a cake or a potato.

So, the task is to use all of the input data to predict the category of object in the picture (cat, cake or potato). A neural network model like this could have millions of inputs, one for each pixel element, thousands upon thousands of neurons spread across a dozen or more layers, and 3 outputs, as shown in Figure 9.

Once the training algorithm has finished and a neural network model exists it can be put to use. If you present the network with a completely new image, such as photo of a potato that you've just taken on your phone, then the model will process the pixels in the image through the network and produce 3 outputs. Output 1 is the probability of the image being a cat, Output 2 a cake and Output 3 a potato.

To decide which image has been presented, you simply compare the 3 outputs and select the one that has the highest probability. If output 1 gives a 5% chance that the image is a cat, output 2 a 15% chance that it's a cake and output 3 an 80% chance that it's a potato, then output 3 is selected. The network model has concluded that the image contains a picture of a potato.

If we now want a more general system, that can identify all sorts of everyday objects, then the same principles apply. It's just a case of having enough images available to train the network, and ensuring that the network is appropriately structured; i.e. the network has a suitable number of layers, and each layer contains enough neurons so that the collection of weights is sufficient to represent the information required to identify all of the different objects.

Figure 9. A neural network model for object recognition.

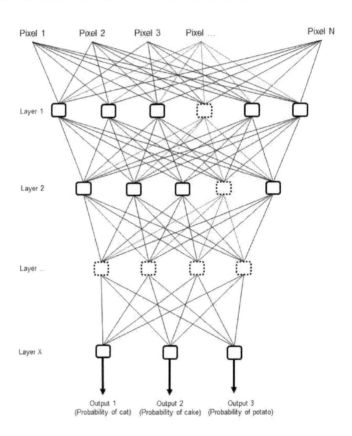

As well as expanding the number of neurons and layers in a network, other avenues of research associated with deep learning consider how the neurons in the network are connected. In a standard network, as in Figures 7, 8 and 9, all of the neurons in each layer are connected to all of the neurons in the next layer. However, other configurations are possible. For example, not connecting all inputs to all of the neurons in the first layer (a **convoluted neural network**). This helps when one is dealing with very large neural networks that require huge amounts of processing power to train, even with the most powerful computer networks available today. As a rule, the more connections a neural network has, the more computer power is required to train it.

Convoluted neural network models have been shown to work particularly well for certain types of problem where you actually don't need all of the input data to produce a very good solution. Object recognition is a prime example of where convoluted neural networks have shown their worth[54]. For images, the interesting stuff is usually in the middle of the image, not at the edges. Therefore, you can perhaps miss out some of the pixels or connections relating to these parts of the picture and hence speed up training time.

Another variant on standard neural networks is to create feedback loops such that the outputs of neurons in later layers act as inputs to earlier layers (a **recurrent neural network**). This makes it possible to incorporate latency, to provide a representation of time or event ordering, which is not captured by traditional machine learning approaches. Examples of where recurrent neural networks are used is in text prediction (like you get on your phone) and language translation tools such as Google translate. The order of the words entered so far are a key factor in predicting what word comes next. What this means is that when you want to predict several words ahead, to complete a sentence for example, then when the network is predicting say, what the third word ahead is, it will use its predictions for the first and second words as inputs.

One of the most exciting recent advances in neural networks is the development of **General Adversarial Networks** or GANs for short. The idea behind GANs is to use two or more neural network models in competition with each other, to generate new and original content[55] that can be passed off as something real, not something that is computer generated. To demonstrate what GANs can do, let's talk through an example.

Let's say, that we want to produce a neural network that can create artificial pictures of people. These are images that look just like real people, but the people in the images don't actually exist. Why would you want to do this? Well, there are lots of reasons. Maybe it's just for a bit of fun, but it might be that you want to use the pictures in advertising or on virtual catwalks. This is so that you don't need to pay human models or get the permission of real people to use their images and so on.

To do this, we start with two untrained neural networks models. For arguments sake, let's call the first neural network "The Judge." And the second "The Student."

The purpose of the Judge is to assess images it is presented with and decide if the images contain pictures of real people or something else. The Judge has a single output that represents the probability that the image is of a real person. If the model output produced by the Judge is more than 50%, then we take that to mean that the judge thinks that the image is most likely to be real. Otherwise, if the probability is less than 50%, the Judge deems the image to be artificial. In principle, the way the Judge works is just like any of the other models we've discussed previously.

Now, let's think about the Student. The aim of the Student is to create pictures of people that look real. The Student is deemed to be doing its job well if it can fool the Judge; i.e. it creates a fake picture that the Judge assigns a high probability of it being a picture of a real person.

The slightly strange thing about the Student model is that it's sort of back-to-front. With a normal object recognition system, the inputs to the model would be a set of pixels, and the output a set of probabilities. With the Student however, it's almost the other way around. The outputs from the model are a set of pixels. The inputs are a set of random numbers[56].

To get things moving, we give the Judge a head start. We train it to identify people using lots of different images that are labeled to indicate if they are of real people or something else. Once this initial training is complete, then the Judge is pretty good at discriminating between images of real people and images of other things.

Now, returning to the second neural network, the Student. The weights of this model are initially chosen at random. The result? A set of meaningless images of random dots. We now present these images to the Judge. The Judge runs through the images and generates a probability of each one being real. Assuming that the Judge has been well trained, it has no problem giving all the images generated by the Student a very low probability of being real – it has no problem identifying them as fake.

The probabilities generated by the Judge are then fed back to the Student. The training algorithm then adjust the weights in the model so as to produce a better set of images, that look more like people next time round. The training algorithm used to train the Judge is also rerun. However, this time, the training data also includes the images generated by the Student.

This process is repeated many many times, with each network continually learning from the outputs of the other. Eventually, a status quo is reached. The Student is now the master. It produces images as good as those contained in the original training data. As far as the Judge can tell, they are real, even though they are completely artificial. In this case, the result is a set of images that look like pictures of real people, but the people simply don't exist. This is how the images on the thispersondoesnotexist[57] site are created.

Creative opportunities for GANs are vast. Train one using pictures of great artists, and images in the style of those artists result! You can then use these for whatever purpose you like, without worrying about copyright infringement[58]. Amazon, for example, has been reported to be using GANs to design garments by using images of fashionable clothing items to train it to produce new designs that have similar stylistic features but which are completely original[59].

Likewise, for music and other creative works. Train a GAN with a recording of a person's speech, and you get something that can speak in that person's voice, which is pretty creepy. Several companies are using this to create audiobooks that speak in the voice of the author or a celebrity, without actually needing these people to perform the narration[60]. Film translation is another area of great potential. GANs can be used to adjust the images of actors speaking, so that the actors look like they are speaking in the translated language, not the original one that they spoke when making the film.

The applications of GANs show huge promise, but the potential for misuse of the technology is also extensive. In particular, it is driving many of the "Deepfake" stories that you may have heard about in the media, where they are being used to create media recordings of politicians and other famous people, showing them saying things that they never actually said. They used to say that a picture never lies, but not anymore.

The most complex neural network models in use today, such as the ones developed by Google's DeepMind subsidiary to beat the world's best chess and Go players[61], combine the features of these different types of neural networks. They have millions of artificial neurons, consisting of hundreds of millions of different weights, connected across dozens of layers.

Although the developers of these networks don't know precisely what each of the individual weights in all those neurons mean, collectively the weights provide a representation of the game and how any given move is likely to impact the game's outcome.

So, why are these types of models so good at what they do? How is it that they manage to capture the nuances of complex problems leading to better judgements than the world's best human experts? One way to think about this is that learning to play a game like chess or Go is analogous to mapping an unknown landscape that the training algorithms explores as it goes. The model weights capture all the features of the landscape and how one gets from one part of the landscape to another using different actions (game moves).

What the data scientists have found is that in mapping the features of a game, or other complex tasks, the training algorithms find parts of the landscape that people have never explored before as well as new (better or more efficient) routes for travelling between one point in the landscape and another. In effect, they have discovered new styles of play that have never been seen by human players before.

Some have argued that in finding these new styles of play the neural networks have displayed non-human i.e. alien, intelligence. However, another perspective is that it's not a new way of thinking, but rather the training algorithms have found new features and patterns that humans have just not got around to thinking about yet. It doesn't mean that the network is necessarily displaying a different type of thinking altogether, but instead, the training algorithm has simply found a new area of the landscape to explore that people have not yet reached.

This principle of mapping and exploration also applies in other application domains, not just games. After training, the different weights in the model contain a comprehensive representation of the problem at hand, and how different stimuli (different inputs) relate to different outcomes (model outputs). One example of the practical application of this exploration approach is supporting scientific discovery. This has proven particularly beneficial in areas of research where huge numbers of possibilities have to be explored to find useful solutions – needle in haystack type problems.

A classic example of a needle in a haystack problem is predicting the 3D structures of complex molecules such as proteins. This is important because knowing the structures of these molecules gives an insight into how they could be used for all sorts of medicines and other useful chemical compounds[62]. There are literally trillions upon trillions of possible protein molecules, but only a handful of these have any practical applications. Being able to determine the likely shape of a molecule from its component parts means it is much easier to identify which molecules are likely to be useful and which are not.

The training algorithms required to create the gigantic neural networks used for things like object recognition, language translation and molecule structure prediction require huge amounts of computing power. It can take days or weeks to complete the training process, even when hundreds of high-end servers (computers) are dedicated to the task 24 hours a day. However, from a usage perspective, that hardly matters. Why? Because once the training algorithm has completed its task, the implementation of the resulting model is comparatively easy.

Very little resource is required each time the model is used. If it's good enough, then I only need one object recognition model. If the model can identify all the objects I need it to, then I don't need to run the training algorithm again and again.

These days, many apps allow you to search your photo library using text. If you search for "Cat" then all of your pictures will be submitted to an object recognition model. In cases where the model predicts it sees a cat, these photos will be presented to you in a fraction of a second. A huge amount of resources may have been required to develop the model that can do that, but implementing the model is much easier.

What's amazing about machine learning models, and deep neural network models in particular, is that they can be trained to predict almost any type of outcome you can imagine to a degree that equals or exceeds the abilities of the best human decision makers *if* sufficient good quality information (training data) is available.

In practically every domain where humans make judgements, either subconsciously or based on their reasoned expert opinion, it has been shown, time and time again, that machine learning models can be built that will do it better; i.e. predict the correct outcomes more accurately than experts can. Want to predict the outcome of baseball or soccer games? Then feed the training algorithm with enough data about historic games and a well-designed model will be better at predicting results than the best human pundits. Want a tool that decides which stocks to buy? then better to trust a machine learning algorithm than a human stock broker. Want to optimize the layout of a factory or supermarket? Then there are neural network algorithms that can do that far better than any human could. I could go on and on (but I won't).

More broadly, and thinking out to the long term (not next year or even next decade, so don't worry yet) the human race only ever needs to develop a super-human AI once. The model(s) that underpin it can be copied millions of times, meaning it's potentially game over for us if that super-human AI isn't on our side.

A model on its own doesn't do much. As I say, it's like a brain without a body. To turn it into a useful application it needs to be deployed into the environment where it can be put to use. A typical AI app will therefore contain the elements shown in Figure 10.

Figure 10. Model deployment.

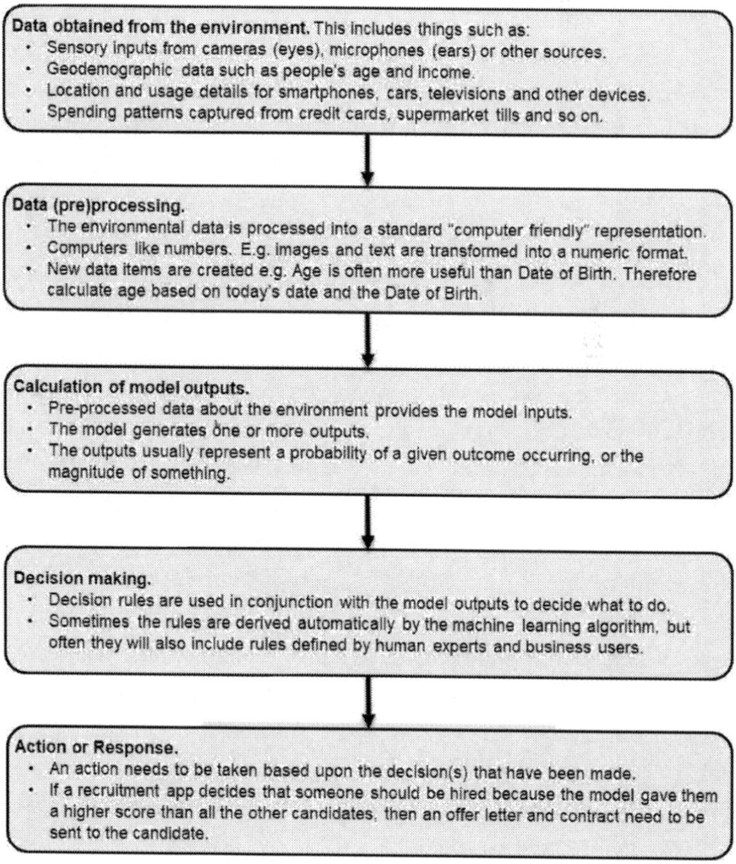

Data obtained from the environment. This includes things such as:
- Sensory inputs from cameras (eyes), microphones (ears) or other sources.
- Geodemographic data such as people's age and income.
- Location and usage details for smartphones, cars, televisions and other devices.
- Spending patterns captured from credit cards, supermarket tills and so on.

Data (pre)processing.
- The environmental data is processed into a standard "computer friendly" representation.
- Computers like numbers. E.g. images and text are transformed into a numeric format.
- New data items are created e.g. Age is often more useful than Date of Birth. Therefore calculate age based on today's date and the Date of Birth.

Calculation of model outputs.
- Pre-processed data about the environment provides the model inputs.
- The model generates one or more outputs.
- The outputs usually represent a probability of a given outcome occurring, or the magnitude of something.

Decision making.
- Decision rules are used in conjunction with the model outputs to decide what to do.
- Sometimes the rules are derived automatically by the machine learning algorithm, but often they will also include rules defined by human experts and business users.

Action or Response.
- An action needs to be taken based upon the decision(s) that have been made.
- If a recruitment app decides that someone should be hired because the model gave them a higher score than all the other candidates, then an offer letter and contract need to be sent to the candidate.

Regardless of the type of model being used, be it a decision tree, scorecard, neural network or something else, it's when the model is combined with these other elements that the model becomes usable and an intelligent tool is created.

What makes some AI applications appear so clever is the sheer complexity of the models and decision rules that underpin them, combined with a slick user interface to gather data and deliver the required responses in a human-friendly way. Combine these components with the latest generation of industrial robots, or integrate them into cars and other vehicles, and one has machines that can interact with their environment and engage with us in a very human-like way.

Let's revisit some of our previous model examples, starting with the diabetes decision tree model (Figure 3). This can be turned into a clever app by integrating it into the GPs systems to be able to use patient records as its input and produce doctor's appointments for the relevant patients as outputs.

For the credit scoring model (Figure 4), this needs be linked to the lender's website/app and be supplied with information about loan applications to generate scores. The scores are then used to decide whose loan applications should be granted. Other rules will then come into play as to what interest rates to charge on each loan and whether or not to try and cross-sell other products and services at the same time. Legal rules will be deployed to ensure that applications from children and bankrupts[63] are automatically declined. Finally, communications need to be sent to the customer informing them of the decision, and the relevant funds transferred to the bank accounts of those applicants who are successful.

When we link an object recognition model to a camera enabled supermarket checkout and the supermarket's price list, then that provides the core components of an automated teller that can see what items a customer is buying. The system can then be programmed to look up the price of that item and issue a statement to the customer telling them what they owe.

As another example of model deployment, let's say that a phone company wants to automate some of its call center activities to save costs; i.e. develop a form of chatbot to answer some customer calls automatically without needing a real person to talk to the customer. What you tend to find in most call centers, regardless of what the organization does, is that most customer calls cover just a handful of standard questions. If you can identify when a customer is asking about these things then you can potentially automate the responses. This leaves your human staff to deal with the less common customer queries that are much trickier to answer.

When a customer calls, they are greeted with a suitable automated message, such as: "Hi there, how can I help you today?" When the customer speaks, several different types of model are employed. First, a speech recognition tool will be used to identify the words and phrases that are being spoken by the customer. This will be very similar in principle to the object recognition model. Instead of pixels, the inputs will be sounds segments, and the model outputs represent the probability of different words and phrases instead of objects.

The words and phrases that are identified then provide the inputs to a second model. This second model predicts the probability of different things that the customer could be calling about. For example, a simple chatbot might have just 4 or 5 outputs. Each output represents the probability that the customer is calling about one specific topic such as:

- **Output 1**. The probability that the customer is asking about their remaining data.

- **Output 2**. The probability that the customer is asking about their remaining minutes.

- **Output 3**. The probability that they are calling to report their phone as lost or stolen.

- **Output 4**. The probability that they are calling to change or cancel their contract.

- **Output 5**. The probability that they are calling about something else (but we can't determine what).

Based on what the first model believes the customer said, this second model generates probability for each of the five outputs. Again, this is just like the object recognition system. The output that has the highest probability is chosen as the one that the model thinks is most likely.

If the customer is using lots of words like "Lost", "Misplaced", "Theft" and so on, then these will all lead to the model giving output 3 the highest value. The system then plays back a confirmatory message along the lines of: "OK, I think what you are telling me is that you have lost your phone, is that correct?" The system then waits for the customer's confirmation. If the answer is yes, the system moves on to ask the customer if they want their phone to be blocked and so on.

Being able to discern only five possible outcomes might not seem very sophisticated. However, even if there are thousands of possible queries that a customer could ask, the vast majority of calls will nearly always be about just a few core issues. Therefore, let your chatbots deal with these and allow your human customer service reps deal with the rest.

Finally, let's consider something a little more sinister and controversial: an AI enabled weapon, trained to track down and kill enemy combatants independently of a human operator. In other words, a killer drone. One approach is to train an object recognition model with thousands of pictures taken from combat zones. The various different features in the images are marked as "Friend" or "Foe" and the system learns to identify which is which.

The model is then integrated into the drone's targeting system that also includes cameras, maps and GPS. The drone is then set free to search a given area. As it goes, it takes pictures of everything it sees and identifies objects in the images. If the system predicts that a certain object is a "Foe" with say, more than 99% probability, then it fires off its rockets or whatever and moves on to the next thing it sees – no humans involved.

5. A Bit Further down the Rabbit Hole…

"…there's a proverb which says 'To err is human,' but a human error is nothing to what a computer can do if it tries."[64]

Well done in getting this far. If you managed to get through the last two chapters then I'm sure you'll be glad to know that that covers most of the technical stuff. Just this chapter to go (and it's quite a short one). We can then move on to talk about some other things.

All of the different ways of building models that we've discussed so far have assumed that there is a database which contains data about the environment, which includes details about the thing you want to infer (predict). If we think about the three tasks we've considered so far:

- In Chapter 3, when building the decision tree model (Figure 3) to predict diabetes, we had data about patients' medical history at a given point in time, together with labels to indicate which patients subsequently went on to develop diabetes and which did not.

- In Chapters 3 & 4, when creating the scorecard and neural network models for credit scoring (Figures 4, 7 and 8), we had information about the applicant, plus a label to show if they subsequently repaid their loan or not.

- In Chapter 4, when talking about using a deep neural network to identify images of cats, cakes and potatoes (Figure 9), each of the thousands of images in the training data was labeled to indicate what was in the picture.

For all of these tasks, each record used in the training process had a corresponding label to indicate what the outcome was. Machine learning training algorithms applied to labeled data; where each case in the training data has both observation and outcome data, is referred to as **supervised learning**.

The vast majority of AI/machine learning applications that you will come across in the everyday world (whether you know it or not), such as target marketing, voice recognition, fraud detection, content recommendation and employee vetting will be examples of supervised learning. If you have a collection of labeled data available and you want to use this to predict some type of outcome or event, then a supervised learning approach is usually the right one to follow and will yield good results.

There are however, certain types of activity where there isn't a database of labelled data for the training algorithm to use. When labelled data is not available, a different set of techniques, referred to as **unsupervised learning,** can be applied.

Unsupervised learning approaches are great at helping to identify and group things together based on similarities between them, but they don't provide you with predictions that tell you how someone or something is going to behave. The objectives and outputs from supervised and unsupervised learning, are therefore, very different. With supervised learning, the end game is almost always a predictive model of some sort. The model can then be used to predict the behaviour of new cases when they present themselves – as we have seen with the credit scoring, diabetes prediction and object recognition tasks. With unsupervised learning, the output is a representation of the structure of the data which is a different thing altogether.

The most common type of unsupervised learning in use today is called ***clustering***[65]. The goal of clustering is to identify similarities and/or connections within data such that you can group (cluster) similar cases together. The idea is that because cases in a given cluster have a lot of very similar attributes, then you can treat everyone/everything in that cluster in a similar way.

To illustrate clustering in action, consider Jenny, the manager of a bowling alley. Most of her customers come to play at evenings and weekends. This is presumably because they are at work or school the rest of the time. However, in the week, she does open during the daytime, even though the bowling alley is not particularly busy then. What Jenny would like to do is understand more about the types of people who go bowling during the day. She can then target promotional offers at similar people in the local community, who might be interested in bowling at that time, and hopefully, boost the Bowling Alley's revenues.

This is an unsupervised problem because Jenny doesn't have any information about how people have responded to marketing activity previously. She doesn't have a list of people who were sent promotional offers for bowling in the past and a corresponding label indicating if they took up the offer or not. This rules out developing any type of predictive model built using a supervised learning approach.

Jenny does however, know who in the local community has been to the Bowling Alley recently. This is because these days, most people leave a record of their attendance via the payment method they used or location information from their smartphone. Once you have that, then a whole host of other information about those people can be obtained from a variety of sources.

Two pieces of information that she thinks are particularly relevant are the age of bowlers and how far from the Bowling Alley they live. Figure 11 shows a plot of these two items for a random sample of early bird customers.

From Figure 11 two groupings (clusters) are easily apparent. A group of young people who live quite close to the Bowling Alley and older people, predominately of retirement age, who live much further away.

Figure 11. Clustering.

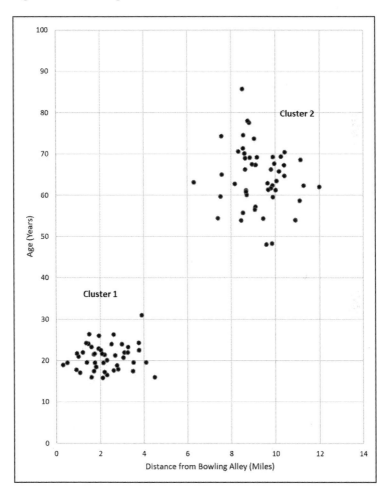

Figure 11 clearly shows that there are very few young customers who live more than a short distance from the Bowling Alley. Also, the age range of those in cluster 1 is quite narrow; it's almost entirely late teens to mid-twenties. The other interesting feature in Figure 11 is that there is almost nobody aged between 30 and 50 visiting the Bowling alley during the day.

Jenny can't be sure as to why the clusters appear in this way, but she hazards a guess that younger bowlers are predominately college students with gaps in their teaching schedules. They may also rely more on public transport, making it less convenient to travel a long way to get to the Bowling Alley. Likewise, the oldies probably have more flexibility in their daily schedules than your typical wage slave, have more time on their hands, have a car and are willing to travel further. They also tend to live in larger houses out in the suburbs, compared to students who tend to live in student accommodation nearer the Bowling Alley.

Although the clustering in Figure 11 is pretty simple, it can help Jenny formulate a marketing strategy tailored to each of the two customer groups. This could be in terms of:

- Travel offers reflecting different types of transport used to get to the Bowling Alley.

- Discounts on extras that are likely to appeal to each group. Maybe soda for younger people and coffee/tea/wine for the seniors.

Equally importantly, Jenny knows that very few mid-day bowlers are middle-aged. Consequently, she should not waste her marketing budget trying to persuade them to come to the Bowling Alley.

The Bowling Alley scenario illustrates the concept of clustering in a very simple way, using just two pieces of data (age and distance), but the same principles can be applied when there are hundreds or thousands of data items available about people. The trouble is it's just not very easy to represent the clusters graphically once you have more than 2 or 3 items of data[66].

There are many commercial products based on clustering.

Experian's Mosaic product assigns all 1.8 million postcodes (zip codes) in the UK[67] to one of 66 clusters. Experian gives each cluster a title such as "Uptown elite", "Low income workers" and "Classic grandparents" to represent the types of people who typically live in those areas. Geo-demographic information about the average age, wealth, number of kids, credit usage and so on is provided for the inhabitants living in the postcodes of each cluster. Organizations then tailor their customer engagement strategy to match the demographics of the people in each postcode. This works because people living in households in the same neighborhood tend to have similar demographics. If you are a highly paid lawyer living in an area classified as "Uptown elite" then it's far more likely that your neighbors are also highly paid professionals as opposed to families surviving on benefits or the minimum wage.

Many clustering algorithms are based on the principle of minimizing the *distance* between cases in the development sample. Distance in this instance doesn't necessarily refer to physical distance, but how two cases differ in terms of specific items of data. The distance between two individuals aged 23 and 25 is less than that between two people aged 17 and 76. If we are talking about smoking habits, then two smokers have a distance of zero whereas a smoker and a non-smoker don't. The distance between two people within the same pay grade is typically less than the distance between two people on different pay grades and so on.

There are many different clustering algorithms available, but one of the most popular is called **K-means clustering.** This is the type of clustering Experian describe using in the design of their Mosaic Product[68].

With clustering, there isn't a model that is produced at the end of the process. All there is, is an identifier to say which of the K clusters an observation has been assigned to. You may be placed in cluster 9, but I on the other hand, have been assigned to cluster 4. When there are new cases that need to be assigned to a cluster, the clustering algorithm can be rerun with the new cases included[69].

Customer profiling applied to marketing is one of the of the best known applications of clustering, but clustering based approaches are also being applied successfully to many other tasks. One example is document clustering. In fields such as law and academic research, there is a requirement to regularly trawl through the ever-growing pile of published literature to find information relating to certain subjects or types of research.

In some subject areas, there are millions of papers and articles about that subject. The mathematics database zbMATH contains about 4 million entries with reviews or abstracts from 3,000 journals and 180,000 books[70]. In medicine, almost a million academic papers are published each year[71]. The traditional way of searching for documents is similar to using Google or Bing – one types keywords into the database's search engine, which finds papers that match those keywords. However, search terms don't necessarily bring back all of the relevant documents and many are returned which are not relevant but contain the keywords in a different context. Some documents, particularly older ones, may not have any keywords associated with them or use different terminology, which means they are missed in the search.

Document clustering is, in essence, no different from the Bowling example we talked about earlier. Instead of geo-demographics; word counts, sentence structure and other document features provide the data items used in the clustering process.

Document clustering can also be applied to tweets, newsfeeds, blog posts and other rapidly changing media in real time. Pictures and video can also be categorized in a similar way. News organizations use these approaches to automatically flag up new posts about specific topics as they appear. This is so that they can include them in their own media publications almost immediately. Social media companies and governments use similar methods (as well as predictive models) to support the identification of illegal or undesirable content.

Supervised learning approaches work very well when there is a large amount of data and each case has a clear unambiguous outcome (label). If there are no labels available, then unsupervised learning can sometimes prove useful; albeit in a somewhat different context. There are however, situations where there may be no observation or outcome data initially, but the learning process is able to assess its performance on a case-by-case basis as it goes. The model is adjusted each time, based on some measure of success or reward, which is calculated each time a task is attempted. This third type of machine learning is called ***reinforcement learning***.

A somewhat over-simplistic example of reinforcement learning is training a neural network to find the highest point in the local terrain using the least number of steps possible. At the start, all one knows is the current map location (longitude, latitude and height above sea level). The network can generate 4 possible scores, each representing a move forward, backwards left or right. The current state of being; i.e. the current location, provides the input (observation data). The resulting state of being, i.e. the new location after the move suggested by the network has been executed, is the outcome.

After each move, the new state of being is assessed. Are we now in a better position (higher above sea level) than before? If the answer is yes, then the algorithm is deemed to have done well - the measure of reward or success is high. If not, then the algorithm is deemed to have performed poorly – the measure of success is low. The algorithm then adjusts the model weights based on the degree of success that its action has resulted in.

A more refined version of this algorithm would also consider past movement history; i.e. where it has been, as part of the input data. It would also make sense to include a longer term view of future success rather than just the one move ahead position. For example, penalizing the success if the same location is visited more than once or providing a reward (increased success) for entering new, unexplored areas of the terrain even if the initial move into that area does not find a higher point initially.

Data scientists will often talk about their being three distinct types of machine learning; i.e. supervised, unsupervised and reinforcement learning. However, reinforcement learning shares a lot of features with supervised learning. As with supervised learning, the algorithms used in reinforcement learning deliver a model, typically based on some form of neural network. Likewise, the model is adjusted based on an assessment of outcomes that results from a given set of input data. As more iterations of the algorithm occur and the weights in the model are refined so the model's performance improves.

The terms "Reward/success" and "Failure/penalty" tend to be applied to reinforcement learning, but these are very similar concepts to "Accuracy" and "Error" that are used to describe how well a predictive model, developed using supervised learning, performs. A small error between actual and predicted values in supervised learning; i.e. a very accurate prediction, is broadly equivalent to a high degree success in reinforcement learning. In the same way, an incorrect classification or inaccurate prediction in supervised learning is analogous to low reward/failure in reinforcement learning.

A great example of the difference between supervised and reinforcement learning is how model training occurs to create a chess playing program. A supervised approach would take thousands of game moves (or sequence of moves) from previously played games as the observation data, with the labeled outcome data providing an indication of if the move was a good one or not. The scores produced by the model are used to indicate which piece to move and to where. The algorithm then finds the model weights that result in the overall best set of moves, measured against the moves contained in the development sample.

With reinforcement learning, no data is provided initially – none at all! The model scores still indicate which move to make just like the supervised approach. However, initially these will be more or less random, given that there is no data to train the model against. Each time a move is made the status of the board (the new state of being) is re-evaluated. The algorithm then adjusts the weights in the model based on how successful its move was deemed to be.

Evaluating how successful a move is in chess is complex and

will often incorporate probable future states of being as well as the current one. However, for the purposes of this example, a simple success criteria is the difference in the value of each players' pieces remaining on the board after a move has been made[72]. If your move results in the taking of a high value piece, but not losing one yourself, then that's a strong success. Making a move and then having one of your pieces taken is a failure, yielding a low measure of success. The ultimate success or failure is losing one's king – checkmate. In this way, by assessing each move and adjusting the model weights accordingly, the program learns by itself without needing to be supplied with any prior information.

A key advantage reinforcement learning has over supervised learning is that there is no limit to the set of moves that can be explored as the algorithm modifies the model weights. With supervised learning, you are limited to the labeled examples available in the training data. For a game like chess, even a huge amount of training data, containing all the moves and outcomes from millions of games, will contain only a very tiny proportion of all possible moves. This was demonstrated very effectively when Google's DeepMind AI team used two reinforcement algorithms to play against each other. Not only did the resulting model outperform the best existing chess program at the time, but during the process the algorithm discovered completely new strategies of play, previously unknown to human grandmasters[73].

Reinforcement learning has generated a lot of excitement because the learning process is very much like the way living beings learn. If I'm trying to learn to juggle then I don't have any information to begin with. I don't have thousands of past examples to hand to learn from. Instead, I try, I fail, I try and fail again, I try and I manage to juggle for couple of seconds, and before you know it I'm juggling like a pro! With each attempt my brain is subconsciously learning more about its environment and refining its actions based on each success or failure; i.e. the length of time I've been able to keep the balls in the air without dropping them.

Reinforcement learning has potential, but it does have its

weakness and is better suited to some tasks than others. One issue is that the training algorithm is bounded by the speed of the trial and error process. If a reinforcement algorithm is being trained to make mortgage lending decisions, then the time between an action being taken and the assessment of how successful that action was could be months or years. Consequently, the training process will take far too long to be of practical use.

Deciding who to lend money too is a far simpler problem than chess and there is usually a lot of labeled development data available. Therefore, you will get better results much more quickly with a supervised approach for tasks such as credit risk assessment. The same issue applies to any type of task where success cannot be measured very quickly. If we return to the chess example, the DeepMind chess playing program needed to play 68 million games to become as good as it did. It could only do this by playing against another computer, allowing games to be played in a fraction of a second. It could not have achieved the same results in a reasonable amount of time if it had been playing against human competitors in the real world. The training process would have been far too slow.

Another weakness of reinforcement learning is the cost of failure during the training process. If you embed a reinforcement learning algorithm into an organization's recruitment policy, then you are going to get a lot of terrible hires initially which is going to cause all sorts of problems. In a similar vein, you wouldn't want to connect a reinforcement learning algorithm directly to the controls of a passenger airliner without some very strict restrictions in place[74].

OK, so finding the tops of fictional hilltops and being superhuman at chess is all well and good, but what are the real-world applications of reinforcement learning? To be honest, not a huge number at the moment. I've done quite a lot of research and my conclusion is that the number of reinforcement learning based solutions in use in world today is pretty small, compared to the number of supervised leaning based ones. Sure, there are lots of research papers and interesting articles about how great reinforcement learning is, but if you go and see what your average organization is doing in their offices and factories today, then you won't find that many examples of reinforcement based systems compared to other types of AI approaches; i.e. supervised and unsupervised learning.

If I had to put a figure on it, I'd guess that more than 95% of real-world application utilize supervised learning and most of the remaining 5% are unsupervised approaches based on clustering. However, I expect that these proportions will change over time as the technology matures. Intuitively, if you have algorithms that can build something that plays poker, chess or Go better than any living person, then they must have potential in other areas.

One area where reinforcement learning is having an impact, is in improving the efficiency of sophisticated control systems in uncertain or chaotic environments. Complex systems, such as power grids, heating systems and **server farms**, have lots of controls which are adjusted to manage different parts the system. The relationship between the controls and the performance of the system is not always obvious. It's not a simple case of one control impacts just one thing – everything is interconnected. Tweaking the water pressure in the cooling system of a power plant to improve turbine performance has a negative impact on the efficiency of a transformer further down the line. Just like the chess problem, where there is an almost infinite number of possible games that can be played, in a power plant there are an almost infinite number of combinations of control settings. Reinforcement learning finds better ways of setting the controls through trial and error, to optimize the overall efficiency of the system. This mirrors the way an experienced engineer would use their knowledge and intuition, learnt over many years of practice, as to what the system wide effects of tweaking different controls are likely to be.

Another area where reinforcement learning is showing promise is in robotics. This is to train robots to carry out complex manual task that could previously only be undertaken by a trained person. A robotic device is given the task of doing something like flipping burgers, sorting items or stacking shelves. Through a process of trial and error they can potentially learn to do this very effectively.

6. AI at Work

"Logical but not reasonable. Wasn't that the definition of a robot?"[75]

Am I going to lose my job to a robot? That's probably one of the most common questions people post when they hear news stories about the way artificial intelligence is revolutionizing the workplace.

Isaac Asimov famously explored this theme in his dystopian sci-fi detective novel "The Caves of Steel." A theme running through the book is that ever more human-like robots are taking over more and more everyday jobs from people. There is a poignant moment when the policeman hero realizes that even his specialist job, as a detective investigating a murder case, may be under threat from an android detective. However, as the story develops and in the sequels that followed, it becomes apparent that both human and robot brings something to the party. The crime is solved through a partnership between the mechanical and the biological. Neither would have solved the murder on their own.

Asimov was quite prescient in his thinking when writing this book in capturing two of the most common themes that tend to arise when we talk about artificial intelligence and how it may impact us in the workplace:

- **Automation.** The replacement of people by computers, robots or other machines, making those people who used to do the job redundant.

- **Augmentation.** The machines provide additional capability, allowing people to do their jobs better than they did before.

One thing to learn from previous technological advancements is that everything doesn't happen instantly, all over the world at the same time and to the same degree. Just because a technology is useful, or is better than what already exists, doesn't mean that universal adoption of that technology is immediate or inevitable.

In the US, pretty much all new cars sold these days are automatics. In the rest of the world however, it's not far off a 50/50 split between automatic and manual gear shifts. There is very little clear logic to this. Often, it's the same cars with the same engines and near identical performance and fuel economy – but buyers just choose to buy manual versions. That's what they're used to and what they prefer. To be fair, automatic transmission sales are increasing year on year worldwide, but I expect it will be many years before manual gear shifts disappear completely from all new cars.

In a similar vein, if you live in North America or Europe, you may think that pretty much everyone has access to the internet, except maybe some gun toting weirdy beardy types living off grid in the wilderness somewhere – I mean, that's just basic stuff like running water, right[76]? But in reality, today, almost half the world's population has no access to the internet at all[77] while most of the other half has taken it for granted for years.

Bearing this in mind, when it comes to automation and augmentation, the big questions that people should ask are:

- How much of each will occur?
- What professions will be impacted and to what degree?
- When and over what timescales?
- Where will they occur, in what countries and regions?

If our AI-driven future is all about augmentation, then that's great. This is because everything we do we'll do better than before, improving outcomes for businesses, customers, ourselves and society at large. However, if the majority of jobs are automated in say, the next decade or so, then that's going to cause all sorts of problems across society. At a simplistic level, if no one has a job then no one can buy anything and the economy collapses. It doesn't matter that a robot can stock the shelves of your local store for a fraction of the cost of a human worker if there is no one who can afford to buy what's on the shelves.

In having this discussion, one thing to bear in mind is that AI can be used to support automation, but just because something can be automated with a sophisticated machine that's not the same thing as an AI powered machine. A huge amount of automation can, and already has, occurred without any AI being involved. If it's a case of sticking labels on a box on a production line more quickly and efficiently than the people currently doing that task, then there are lots of mechanisms that can be used to do this automatically, without them needing to be intelligent.

Having said that, AI is driving all sorts of intelligent automation but opinions differ as to what the impact is going to be, in what occupations and how quickly we will see those changes occur. Some argue joblessness will be through the roof in just a few years' time. Others expect a slower and more gradual change that will allow time for the workforce to adapt. As at today, it's difficult to know which view will prevail.

Automation is obviously worrying from an individual perspective because of its potential impact on people's livelihoods. So, let's start by looking at where and how automation is occurring and thinking about what the impacts of automation are likely to be. Politically, these sorts of questions are also very important and the repercussions of AI-driven automation will no doubt have a political impact just as much as economic and social ones. The following is an (abridged) extract from a speech given by the soon to be Prime Minister of the UK about the challenges presented by the coming age of automation[78].

> The danger, as things are, is that an unregulated private enterprise economy will promote just enough automation to create serious unemployment...
>
> Let us look at what is happening in automation all over the world... in the United States they have reached a point where a programme-controlled machine tool line can produce an entire motor car...without the application of human skill or effort. They can do this without a single worker touching it...
>
> The essence of modern automation is that it replaces the hitherto unique human functions of memory and judgement. And now, the computers have reached the point where they command facilities of memory and of judgement far beyond the capacity of any human being or group of human beings who have ever lived...
>
> In America technological change is beginning to move now ever more rapidly in the white-collar professions...and let us be clear that in America today and in Britain tomorrow we face massive redundancies in office work no less than in industry...
>
> And when machine tools have acquired, as they have now, the faculty of unassisted reproduction, you have reached a point of no return where if man is not going to assert his control over machines, the machines are going to assert their control over man...

The same things that the future Prime Minister said in his speech are being reiterated in the media all the time. Giant unregulated corporations driving joblessness, highly automated factories, shops and offices replacing professionals as well as the working classes, all who will be thrown on the scrapheap of history – not to mention machines that can build themselves!

In fact, it's all so worrying that I'm just waiting for my robot replacement to walk through the door and shake my hand before I'm escorted off the premises, leaving it to finish off the rest of this book. It will probably do a better job than me anyway.

How much of what Harold Wilson said will come to pass? Who? Harold Wilson. Who? The British Prime Minster 1964-1970 and 1974-1976. This speech was made in 1963, many decades ago, shortly before he was first elected to office.

Was Wilson right, wrong or just ahead of his time? Wilson was living in a very different time and it can be argued that he and the wider world didn't really have a full understanding of what AI and machine learning could achieve back then, and even if they did, they didn't have the technology to deliver it. However, regardless of the state of play back in the 1960s, what Wilson actually says echos, almost word for word, the same concerns that are being raised today about the potential disruption that AI-driven automation may cause, driven by giant, uncaring and poorly regulated corporations whose only concern is their stock price. In particular, it's not just about the automation of simple production line tasks, but about machines with intellectual capabilities that can replace all sorts of white-collar jobs traditionally done by university graduates with years of experience and specialist training.

Wilson wasn't wrong. First world economies underwent huge technological upheavals in the final decades of the 20th century. The near death of the loan underwriting profession, that's been a theme throughout this book, is just one of many professions that have all but disappeared due to technological advances since Wilson's time. TV repairers, switchboard operators, film projectionists and typesetters are just a few examples of jobs that effectively no longer exist due to advances in technology. Similarly, the automation of manufacturing processes across swathes of industry in the 1980s and

1990s, was a key factor in the decimation of the workforce in those industries during that time[79].

If we take a look at another major industry sector, farming, then in the middle of the 19th Century, around 22% of the UK population was employed in agriculture[80]. By the start of the 21st Century, this figure was below 1% A massive reduction driven by a mixture of evolution and revolution in farming methods. This included the introduction of ever more complex machinery to replace tasks such as sowing, reaping and threshing that were once mainstay activities of the farming process[81]. If we are talking about AI automating farming, then you could argue that it's a bit late for that – most of the jobs have already gone. Sure, artificial intelligence may reduce that 1% figure a bit more, but it's very much the icing on the cake rather than the cake itself.

The message I am trying to get across is that change and automation has been a feature of advanced economies for decades, if not centuries. Concerns over job security and the impact of technology on society is not new either. Arguably, it's just that this time round we have a different sort of technology, AI, added into the mix.

If you think that that argument holds water, then asking if more jobs are going to be automated in the future due to AI (or other technologies) isn't really the right question to ask – the question has only one answer – of course some jobs will be lost due to AI. This is because it has been demonstrated beyond doubt that artificial intelligence can facilitate doing many things better, quicker and cheaper than a real person can in many industries and professions. A much better questions to ask is:

> Are AI technologies so different, that they will result in an unprecedented acceleration in automation and rising unemployment, the likes of which we have never seen before?

If AI is just a means of continuing current trends in automation then no problem, it's just business as usual – no point getting hot under the collar about it. Some types of job will become redundant and disappear, some will adapt and change. Others will see more people employed in those sectors, and in some areas, completely new types of job will be created.

If, on the other hand, the rate of change is going to increase significantly because artificial intelligence brings something else to the table, then that is a concern and that is what's likely to cause problems to both individuals and societies. The important factor is the balance between job losses in one sector and the rate that new jobs are created in another.

So, which way will things go? It's difficult to tell, but let's look at what some experts in the field have said and also examine some trends that are being seen in the economy.

A very influential study of its time, back in 2013[82], estimated that 47% of jobs in the US were at high risk of being lost due to automation by about 2035[83]. The figures for other first world economies such as those of Germany and the UK were similar. That's pretty scary stuff, and will be very worrying to a lot of people, but things have mellowed somewhat since then. More recent studies have revised this figure down. A survey by PwC in the UK a few years later estimated an upper limit for the number of job losses of 38%[84] which they later revised down to 20%[85] PwC also concluded that jobs losses would be offset by new jobs that would be created. This aligns reasonably well with recent figures from the OECD.[86] These estimated that only around 14% of jobs might go. The OECD also acknowledged that this figure was only looking at job losses and didn't take into account all of the extra jobs that were expected to be created.

This seems to be where we are at today; i.e. no more than about 10-15% of jobs will go in the short to medium term (next 10 – 20 years) due to AI-driven automation. That's still a pretty big impact, but a lot smaller than what was originally being talked about. Also, we are not talking about these things occurring overnight. If you spread those 15% jobs losses over a 20-year period, then the picture looks even better for the poor worker. It equates to just 1 in every 130 jobs lost each year, and this is not taking into account any new jobs that will be created in tech industries. A pragmatic way to think about this is say, you work in a team of 20 people, then over the next 10-20 years, that may reduce to 17 or 18, all other things being equal. Even if you take the worst case scenario of half of all jobs disappearing in the next 20 years, then that's still only an average of 1 in 40 jobs each year.

You also have to bear in mind that a lot of these studies are talking about what could theoretically be automated, not what will be. All large organizations I know are riddled with bureaucracy and waste. Very few operate at anything that comes close to 100% efficiency. Just one example of this is where, for a company I had dealings with, all of the IT staff had really tiny monitors. Some of these were not even HD resolution. Anyone who works with computers knows that having at least two large high resolution screens is pretty much the norm in the industry, and is essential for maximum productivity. This is because IT staff are often cross referencing between several detailed high-resolution items. It makes sense to be able to compare things side by side, and the more of a document you can see the better. Upgrading their screens would have had a significant productivity impact for quite a small cost. Yet, all the focus from senior management was on automating other parts of the business because AI/automation was what some highly paid consultants had told the CEO that that was what they should focus on.

Another issue is that in some large organizations, and government departments in particular, senior staff can be obsessed with the delivery of big projects, rather than improving business processes and improving profitability. Why? Because in government there is often no bottom line as such. You make your mark and get your promotion from delivering something big that the politicians want to see, not for beating your financial targets or delivering something useful.

These types of behaviours and inefficiencies have always existed in large organizations and always will, but that's not really all that important. All you have to do to be successful in business is be a little bit faster, a little bit cheaper and a little bit more effective than your competitors. Do that and you'll win the race.

It's a bit like that old joke about two people in the jungle being chased by a hungry lion. One of them stops to put on their running shoes. "Why are you doing that?", askes the other one. "No matter how fast you run, you'll never be able to outrun the lion." To which the first person replies. "I don't have to outrun the lion, all I have to do is outrun you!"

It's good to consider what might happen in the future, as reported in the aforementioned surveys and studies, but let's not forget what's already happened. The past informs the future, and we'd be remiss not to consider how things have been going to date in weighing up what may be to come. It was probably sometime around 2010[87] when the current cycle of concern over the impacts of new technology, and machine learning and AI tech in particular, began to be voiced. So, we've got a good few years of this new technology behind us. Given all the great things we are being told this tech is being used to automate, from driverless vehicles to fully automated warehouses to robotic customer service reps, I would expect that we should be seeing their impact on the economy by now.

One very obvious thing to look at, to see if this is the case, is unemployment rates. These should be rising – surely? Figure 12 shows the pattern of unemployment in the US[88] over the fifty-year period between 1970 and the end of 2019[89].

Figure 12 is quite up and down over time. The troughs correlate with periods of recession in the United States at that time, the peaks with the boom times. You can clearly see the rise in unemployment during the recession following the dot.com bubble of the early 2000s and the fallout from the sub-prime mortgage crisis later in that decade.

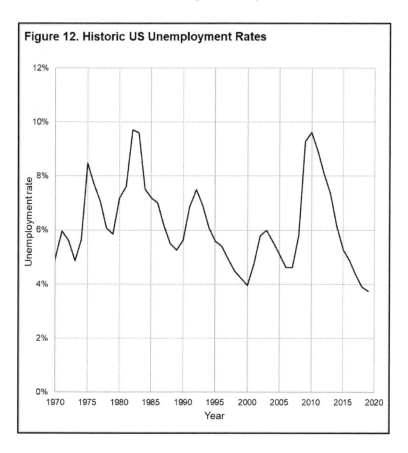

Figure 12. Historic US Unemployment Rates

Over the long term, the graph doesn't show much that supports rising unemployment. If anything, it shows the converse. The fall in unemployment over the last few years is greater than at any time in recent history. If I were to present this data to a classic machine learning algorithm, which just considered correlations between technology and employment, then the model it produced would no doubt conclude that increasing use of AI technologies drive higher levels of employment, not lower ones, which turns conventional wisdom on its head!

Now. I'm not saying that this is the case. You've got to be extremely careful about drawing conclusions from a single data series like this. At best, it's a piece of circumstantial evidence that can be considered alongside any other information available. However, to date, there doesn't seem to have been a noticeable detrimental effect on overall employment levels in the US in the last few years due to artificial intelligence or anything else outside of the normal rise and fall that follows economic cycles. Maybe that will change in future, but we don't have evidence of that yet – which to be honest did surprise me a little.

It may be the case that AI is enabling new forms of employment. It's not just a case of technicians identifying a business process done by people and replacing that process with a robot or a clever computer. Employers are finding new roles and/or retraining those of their staff who are being displaced by automation, and there is some evidence that this is what is happening[90]. Instead of laying off staff whose jobs are replaced by AI, they are taking advantage of the extra resources that automation frees up to do other things.

Another, related expectation, is that artificial intelligence will lead to huge increases in productivity. By productivity, I mean the amount that each worker produces per hour worked. A classic example of improved productivity is the case of automating activities in a warehouse. Instead of having say, 10 pickers walking the aisles to collect items from the shelves, clever robots could do that instead, maybe overseen by a couple of technicians. In which case, productivity has increased fivefold – two workers required now whereas ten were needed before.

As with automation, the big question is not if artificial intelligence will help to drive up productivity – it almost certainly will and has been doing so for many years. Rather, the question is: is AI delivering something new and different to what we've seen before? Will the use of artificial intelligence deliver a step change in what a single worker can do? If AI-driven productivity is really such a game changer, then we would expect to see a blip in the productivity figures in the last few years to support this. Figure 13 shows the historic trend in the change in productivity of US workers[91]; i.e. how much more productive workers are becoming each year.

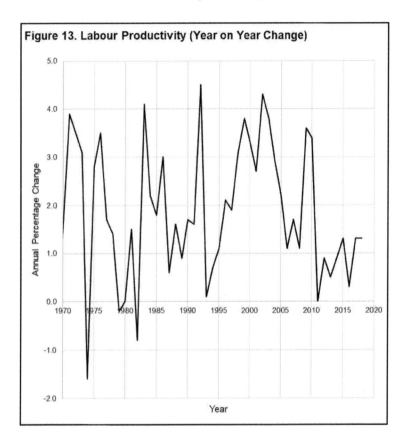

Figure 13. Labour Productivity (Year on Year Change)

What Figure 13 shows us is how much more productive the average employer is than in the year before. In 2015, for example, the figure of 1.3% means that employees produced 1.3% more stuff per hour worked than they did in 2014.

Figure 13 presents a similar picture to the one we saw with employment in Figure 12. The trends are a bit choppy, but it doesn't look like productivity is rising any faster now than it has in the past. If anything, the figures for the last few years are below the long run average. Mmmm.

Again, one needs to be careful drawing firm conclusions from a single graph. There could be all sorts of reasons why the rate of productivity improvement hasn't increased more in the last few years.

One theory is that productivity increases, due to new technology, are being offset by the downsides of other technologies and social changes that interfere with peoples' work. Smartphones in particular, have been blamed for distracting workers and reducing how much work they do as they continually interrupt what they are doing to respond to Instagram, tweets, Facebook posts and the like. The "Always on" culture is also linked to reduced creativity, decimating those periods of quite reflection that can spark new and original ideas. Creativity flourishes when it is given time and space to do so away from constant distractions such as those generated by our various devices. Maybe it's no coincidence that the rise of the smartphone coincides almost exactly with the boom of AI technologies in the workplace. Maybe, in terms of productivity, the two things have cancelled each other out.

Or maybe we've utilized the improved productivity in different ways. Rather than increasing output, it has made the workplace better for workers. Many organizations now offer far more flexible working arrangements than they did a few years ago and some now adopt a "Just let the workers get on with it" attitude – as long as they deliver, who cares how they do that. Go back twenty or thirty years and any employer with that sort of attitude would have been considered crazy.

Another possibility is that a lot of time companies are engaged in the red queen's race[92]. You have to invest and improve, just to keep up. A few years ago all I needed to succeed as an on-line retailer was a nice website that accepted credit card payments, but hey, now I need to build real-time individually tailored customer interaction into my site, apply sentiment analysis to assess what people think about it and chatbots to keep customers interested and engaged. I also need to have relationships with Paypal, Apple Pay, etc., rather than just a single credit card company. And all that's just to keep up with the competition without doing any of the real cutting edge stuff. My internet marketing team isn't dispensed with, they are just refocused on the next set of tasks in hand in the never-ending race against my rivals. If anything, maybe I need even more people just to keep on top of everything.

A different perspective on this productivity puzzle is that it may be the case that almost all medium to large organizations are talking-up the benefits of AI-driven automation, but to date, few have really understood how to apply AI effectively. They haven't been able to fully integrate artificial intelligence into their business processes and strategically realign their organizations to reap the benefits that these new technologies are offering. This is because they don't really understand how the technology works, and therefore, the associated risks and pitfalls they must avoid to get it working as intended.

The taboo, that no board room exec shall ever mention, is failure. A very considerable proportion of the projects undertaken by businesses to apply artificial intelligence have failed or delivered less than expected. This is because they have not approached things in the right way. You'll always hear the head of the business, the CEO, telling everyone about their successes but you'll rarely hear them stand up and tell a crowded room about their failures. I spent a whole chapter on this issue, of why these types of project fail, in one of my previous books in 2014[93]. More recent studies have all put the failure rate for business AI projects at well over 50% i.e. most fail, with the most pessimistic commentators estimating that almost 9 out of 10 business have seen their AI initiatives fail to fully deliver[94].

The easiest mistake for organizations to make, and one that I have come across on more than one occasion, is to hire a load of those mysterious data scientist types without really understanding what you want them for. The data scientists turn up at the office, with more PhDs than you can count, and then apply all of their technical wizardry to problems and processes that they have no real understanding of. Just like any other technology, artificial intelligence needs to be shaped to fit what an organization does. This needs to take into account all of the soft issues involving people, the law, social customs, customer service and reputational impact. Technology on its own isn't enough.

The other big mistake is where organizations try to do it all at once and automate everything they can as quickly as possible. That's a very high-risk strategy. What most organizations should be doing is to start with some simple tasks that need to be improved upon. They then move on to the bigger, more strategic projects, once they have gained a degree of competency with the technology. Many are biting off more than they can chew and then see their projects fail spectacularly.

Big traditional institutions (which covers the vast majority of large businesses and government departments) are often bureaucratic, political and slow moving in nature. It takes time to change things. Maybe we just need to give these organizations a few more years to get their act together and that's when the employee purges will really start to ramp up. It's no surprise that it's the new corporate giants, the Googles, Amazons, Ubers and Facebooks of the world, who have managed to embed AI technologies most effectively into their organizations. This is because they were unburdened by the accumulated drag of organizational cultures and long embedded business practices when they put their AI strategies into operation.

Finally, let's look at wage inflation. This is shown in Figure 14[95]. In a world where cheap robot labor is taking over, I'd expect to see average wages falling as hard-pressed workers are being undercut by their robot competition.

In Figure 14, as with productivity and unemployment, I don't see any recent obvious trends that don't align with known recessions or periods of economic growth. Maybe we just need to wait a few more years to see the impacts of AI, but I don't see the trends that one would expect in the data at the moment.

There may not be a huge amount of evidence for rising unemployment or other impacts at a national level – yet, but some professions are more at risk than others and some have already felt the impacts. Is your job at risk?

Figure 14. Median Annual Wage Growth

Artificial intelligence is most effective at supporting the automation of tasks when:

- Tasks are well defined. You can understand clearly what's involved in doing it.

- There is lots of data about tasks for the AI to learn from.

- There are clear decisions or outcomes that are required, based on the data available for training.

- The tasks are frequent and repeatable.

- A task is measurable. It's easy to determine if the task was performed well, or not.

As we discussed in Chapter 2, current artificial intelligence applications tend to be quite narrow in their capabilities. A typical AI app is very good at specific tasks, but not so great when it comes to general intelligence. When it comes to the low hanging fruit, if you work in an industry where lots of people are doing a lot of similar things, or making certain types of decision on a frequent basis, then these are the roles that are most at risk of being impacted. Even if you think that the task you do is very complex or requires a lot of training, that does not reduce your vulnerability. If what you do is something that can be distilled down to a set of clear-cut actions, decisions and outcomes, then your job is at risk.

The more diverse and one-off the type of work, and the broader the span of someone's responsibilities, the harder it is to develop an AI-based tool to perform that role effectively. That's not to say that an AI won't be developed that can do that role at some point in the future but that it's not feasible now or it can be done now, but the cost is prohibitively expensive, making it not cost effective to replace you yet.

Arguably, there are also some types of job where only a person will do because, it is argued, we like personal interaction and value the human touch more highly than having something that is ruthlessly efficient. Maybe a robot could technically do the role, but we want to keep that 1:1 interaction with a real person. Personally, I like being served in my local pub by a real person. I don't want a machine to pull a pint for me. Maybe we'll start seeing "Real bar staff" signs to entice customers into establishments alongside signs advertising real ale and craft beers. I'm also happy for my lawyer to have artificial intelligence tools to help them give me the best legal advice, but I still want them to give me the advice. I don't want to be talking to a machine. In education, the 1:1 (and 1:many) experience is hard to replace. There are educational robots around that can interact with students to support their education, by asking and answering questions, but they are very crude devices. They are nowhere near being anything like a human teacher and probably won't be for many years to come, if ever.

Others may feel differently about personal interaction. A lot of people don't want that 1:1 experience. In fact, for many mundane tasks, like checking you bank balance or buying a loaf of bread, then having a machine take your money or respond to your question is actually much preferred. This is precisely because you don't have to waste effort to interact with some stranger who you may never encounter again.

If you take all that together, then its things like manufacturing and warehousing, transportation, retail services (such as call centers) and back office functions (routine admin/paperwork) where jobs are most likely to be affected. This is because they involve following a certain set of frequently repeated procedures and behaviours that can be learnt by a computer and acted upon by a robot or other machine. Professional and personal services, education, healthcare and the arts are areas where jobs are less prone to robot takeover[96] - although that's not to say that many functions that support these roles can't also be automated.

As we have been discussing, automation is one consequence of using AI technologies in the workplace and many fears have been voiced that that will mean massive job losses. The other side of the AI coin is augmentation. The argument is that no one needs to lose their job, but the job you do can be done so much more effectively. The key question is not if a profession will be automated, but how much of that profession and what elements.

You'll be able to deliver so much more if AI is deployed to help you. Your job won't disappear, but you are going to need to adapt to doing a different set of tasks in a different way. You'll need to accept that new technology is going to do some of the heavy lifting for you, specifically to free up your time to do all the other more important things that you never spent enough time on before.

The aforementioned study by the OECD come to the conclusion that only about 14% of jobs will be automated in the next few years, but it also said that nearly half of all jobs in developed countries would be impacted by AI in some way, requiring retraining and/or reorientation.

How will AI augmentation help you in your work? Well, a lot of it covers the same principles as automation. It's about having a machine doing things that you might have done before, but quicker and better. However, we are in augmentation territory when two conditions exist:

1. There are still a number of important, significant or time-consuming tasks that the machines can't do very well, or do as effectively as a person.

2. There are lots of extra things that you could be doing, if you had the time. Having some help from an AI will allow you to do those things.

Law is one area that is already seeing the benefits of augmentation. It's a profession where there is huge scope to adopt AI tools more widely to make the whole process of managing cases more efficient.

The law in its entirety covers a huge span of activities and variety of cases. This ranges from high profile murder cases, to fraud, to mass market claims for road traffic accidents, divorce and home insurance. The news has on occasion contained stories of "Robot judges" dispensing justice, and maybe that will one day be the norm, but the key areas of law that are currently being impacted by AI today include:

- **Collating documents.** AI technologies, such as document clustering, can reduce the time spent trawling through the mass of case law to find all the relevant legal documents that might support a case.

- **Conviction prediction.** Given information about a case, what's the chance that the accused will be convicted? A defence lawyer might push for a plea bargain rather than acquittal if their AI helper informs them that the odds are against them.

- **Triage.** AI is used to gauge how difficult a case will be to resolve. A simple damages case gets put in the office junior's in basket, whereas the most complex cases are passed to one of the firm's senior partners to deal with.

- **Claims and damages.** Based on the features of an insurance claim, divorce case etc., tools can be developed to estimate what the value of any claim/settlement should be.

Let's now consider something like education. If you can automate 20% of a teacher's job, does that mean that you can make 20% of teachers redundant? No. Why not? Well, one reason, is that in most schools leaving a classroom unattended is frowned upon. You still need one teacher per class. You can't just leave kids alone in a room with a robot all day. Sure, there are some educational robots around, but these are nowhere near advanced enough to act as substitute for a real person. Instead, in education, the sorts of tools that are being trialed or are on the horizon, that are going to help teachers, are things like:

- **Automatic registration.** Facial recognition scans the class confirming everyone who is there. The teacher only needs to confirm if those the system reports as absent, are actually absent.

- **Marking.** Student scripts and other admin can be automated to a degree.

- **Providing information.** Chatbots can be designed to answer a lot of the questions that parents and students have, without having to bother the teacher.

- **Student support.** Identifying students who may struggle in the future. Using predictive models, that use student information as inputs, it is possible to predict which students will struggle with certain ideas and concepts. Teachers can then work to address student weaknesses *before* they become apparent.

For the private sector, it's very much about the bottom line. Stripping out cost to meet existing demand for your products and services as cheaply and efficiently as possible, which makes automation often the first thing that comes to mind. Sure, you want to increase demand for your products and services, but at the end of the day, there is only so much customer demand for what you've got to offer. There's only so much toilet paper that people are willing to buy. Only so many Pizzas that your customers can consume.

For many charities, government departments and other not-for-profit organizations, there is often a different perspective. Reducing costs and improving efficiency is important but not to improve the bottom line. Rather, it's so that you can deliver more, as much as you possibly can. In health care, the demand for services is almost limitless – no amount of money can provide everyone with the very best in health care. If an artificial intelligence tool can help a doctor diagnose and treat a patient more quickly, then you get the doctor to treat more patients or spend more, better quality, time with patients. You don't reduce the number of doctors. The same principle applies to law enforcement, education and many other public services.

The early AI evangelists were perhaps running before they could walk. They projected current trends into their ideas of the future in simplistic ways. What is now being realized is that although there may come a day when a robot can do pretty much anything we can, that day is probably still many decades away. There are still many technological, legal and cultural hurdles to overcome.

For most of us (but not all of us) that means that our jobs are probably safe from robot takeover for the time being, with one caveat. Be prepared for change. You must be willing and able to adapt and accept that there will be an increasing number of AI-based tools that can enhance what you do, so that you can do your job more effectively and more efficiently than you did before.

7. AI in Society

"I'm sorry, Dave. I'm afraid I can't do that."[97]

These days, artificial intelligence driven technology is being used for all sorts of purposes by businesses, not-for-profit organizations, in industry and by governments. Is AI making society better, worse or just different. In what ways is it doing this?

For some, the big questions are about robot intelligences to whom we risk becoming slaves or pets. Or possibly, some mis-directed super AI that accidently wipes us all out because it's been incorrectly instructed to do something like: "Protect the environment." The all-powerful machine concludes that the best way to do this is to exterminate all the nasty polluting humans and our fate is sealed[98]. Alternatively, a more optimistic view, is that the AI's become benevolent benefactors. Maybe, somewhat similar to the "Minds" found in Ian Bank's culture novels[99]. Infinitely more intelligent than any human but who guide us and provide the support we need to become the best that we can be, in whatever terms we define best to be. They are our helpers, mentors and protectors – albeit our masters and superiors.

I agree that there are concerns that robots may one day decide to wipe us all off the face of the planet or some super-intelligent robot overlord makes us their slaves, but these are not the problems of today, next year or even next decade. The tech is simply not advanced enough and the world is far too fragmented in terms of its industries, politics, national interests and technologies for this type of takeover to occur any time soon.

Today, artificial intelligence is significantly changing the way that factories and industrial processes are undertaken, to deliver products more efficiently and more cheaply. It's also helping to drive forward scientific discovery, improve our understanding of complex environmental issues, produce better consumer devices, improve energy efficiency in our homes and a host of other things. Governments and city authorities are also using AI to improve the quality of public services and the urban environment.

The advances in these areas are not being driven by a central intelligence, but by managers, civil servants, politicians and business leaders. This is to address specific problems and challenges that they have, to meet their targets and beat the competition.

In transportation, we are on the cusp of a revolution in self-driving technology, where we can move goods and people around the country without needing human pilots or drivers, again underpinned by advances in AI. We are already seeing driverless vehicles being used in small well-controlled environments, such as in warehouses and factories, airports and university campuses, and their use is bound to grow in the years to come.

It's going to be at least a few years before we get to a predominantly driverless transportation system but, if we can get there, then that may drive down the prices of many goods and services, boost the economy and give us access to a wider range of consumer goods. An integrated system, where all the vehicles on the roads communicate with each other, can also be expected to massively reduce congestion and pollution because the system would organize itself to be more efficient.

It is also conceivable that we move to a hire only transportation model, where nobody owns their own car anymore. Instead, a driverless vehicle is always on call for the daily commute, to take us on vacation or to transport us wherever else we may want to go. The average person only uses their car for a few hours each week. Therefore, a hire-based model can potentially reduce the costs of ownership, but more importantly, it would massively reduce the environmental impact of car production and car parking. This is because we'd only need a fraction of the total number of cars that we have today, plus we'd also not need to worry about parking them. Once we've got out of a vehicle, it moves on to its next job or finds a quiet street where it can wait until its services are required once more.

Most of the things that AI is being used to deliver are great. In all the areas I've just mentioned, artificial intelligence is helping us to do things better and more efficiently than we could do them before. At the small scale, at a personal level, we are also seeing AI being used to improve our day-to-day experiences in almost everything we do. A lot of these benefits are simple things that most of us are hardly aware of. Things work just a little bit better, are a bit slicker and require less cognitive effort on our part. Whether it's better predictive text, controlling your car via voice command, recommendations from your streaming services, water companies spotting more leaks or more accurate weather forecasts. All of these things are using artificial intelligence to make our lives better. There is an argument that some of these developments are making us lazy, or possibly reducing choice, because we rely on recommendations we receive rather than seeking out new things ourselves, but I think it's hard to argue that, on balance, these things are of overwhelmingly providing a variety of benefits to a large section of society.

That covers quite a lot of the upside but what about the other side of the coin? Today, the most significant ways in which artificial intelligence is being deployed, across almost every industry and government sector, relates to people. Government departments, corporations and other organizations are increasingly using powerful artificial intelligence tools to decide how people are assessed, profiled, manipulated and controlled.

The powers in society (i.e. governments, religions and corporations) have always sought to control and manipulate people in pursuit or their goals. Often, this is to deliver on good intentions to improve the lot of individuals and society, but sometimes it's for selfish ambition with the sole aim of benefiting those organizations and the people who control them. Often, with little regard to the harm that those actions may cause to others. The difference this time around is that the tools organizations can employ are much more powerful and can be targeted more precisely and more accurately than ever before.

In the past, attempts to manipulate populations was often via the use of mass media and through identifying which homogenous group within the population people fell in to. If you are this type of person we'll treat you like this. If you are that type of person, we'll do something else. However, artificial intelligence enables far more granular tools to be deployed. These allow efforts to monitor and control people to be tailored to the specific characteristics and behaviours of each person, instead of much coarser activities aimed at this or that group in society.

These AI-driven approaches are also flexible and adaptable. They can react, and quickly modify the way they deal with you, in response to your changing personal behaviours and preferences. All this is driven by the data that flows from your various smart devices, social media and a variety of other sources, to the server farms controlled by those various powers.

The big concern, as I see it, is that a lot of artificial intelligence is being deployed in an ad-hoc basis such that it slips under the radar. There isn't a single shadowy super organization developing lots of nasty tech or a mad scientist developing a robot dictator that will achieve world domination. Instead, we risk losing control piecemeal to the machines and their masters, one decision at a time[100].

A retailer deploys an AI tool over here to decide who its customers should be. A government automates some services over there to track criminal suspects. A hospital uses an automated assessment tool to decide who should be first in the line for operations. You get fired because some AI selection tool assigns you to the redundancy pool and so on. Before you know it, AI-based decision-making is impacting almost everything you do; at work, in your home and in society at large. This gives the owners of the AI tech enormous power over you. The tech itself isn't deciding to pursue some overarching goal of ultimate control. It's merely a tool, employed by those who control the tech, to control us, to get what they want.

Most people have no idea what can, and is, being done surreptitiously to nudge (or force) them, one way or another, to do what others want them to do. We've already discussed credit scoring and diabetes, but almost any area where decisions are being made about how to treat people has the potential for using artificial intelligence. You just need some data to train the algorithms with and a mechanism to deploy the model that results. Deciding if you should be invited to a job interview, assessing peoples' right to receive state benefits, setting your insurance premiums and deciding if you are who you say you are at the airport, are just a few examples of how prevalent AI-driven technology has become. In all these areas artificial intelligence is being deployed right now, today, to understand, manipulate and control you.

If we continue with credit scoring, then in practice, that's all about the banks making as much money from us as they can. If that means deciding not to lend money to people with certain traits then so be it. From a bank's perspective, all they are focused on is what they can legally do to maximize their bottom line. To hell with people if declining their loan application causes a problem – it's not our concern.

In some countries, government regimes are adopting AI-powered technology to track and monitor entire populations, to detect and predict those most likely to be "Individuals of interest."[101] i.e. people who don't conform to the norms of society that the government dictates. In principle, this is somewhat analogous to credit scoring. Instead of calculating a measure of one's creditworthiness, the algorithm produces a "Social score" that represents your qualities as a citizen. Your score is based on factors such as what job you do, the type of social contacts you have, what you buy, social media posts, political leanings and so on. Score highly, indicating you are a good citizen, and the world is your oyster. You can get a promotion, open a new bank account and all the other things that one expects to be able to do in society. However, if you score at the bottom end of the scale, then that indicates you have undesirable qualities. Your activities will be restricted, and if necessary, intervention taken to bring you back to the straight and narrow so that you conform with the accepted norm. This type of application of artificial intelligence is a potential risk to human rights, restricts freedom of expression and threatens democratic principles.

You might think that this type of individual assessment would only be carried out by oppressive regimes. However, many governments around the world, including those in Europe and North America, are looking to use similar tools to identify "Undesirables" within their populations – however they define that. Many large corporations, including Facebook and Uber, have also been reported to have profiled their customers and employees in a similar way[102]. Only this time, score low enough and you risk being sacked, censored or having your account closed, without being provided with a detailed explanation as to why or what you may have done wrong. Hence, there is little opportunity to challenge the decision that has been made.

Unlike your credit score, which you can ask for by law in many countries, the scores and metrics used by these companies to assess you are largely a secret. They are known only to the relevant departments within those organizations. The scores are then used to decide what course of action should be taken to ensure that their goals, not yours, are achieved. However, as we shall discuss in the next chapter, in many countries legislation is evolving to try and protect consumers and employees against these types of activity.

That's the scary bit. Now, I'm not saying that all, or even most, uses of this type of AI-based control are necessarily wrong – far from it. When a government uses artificial intelligence for social good, to prevent diabetes or other diseases I think that's great. If they use it to identify vulnerable people who need support, or to improve education and so on, then that's potentially a very good thing if executed appropriately[103]. It helps individuals directly, but it also frees up resources to spend on other things that society needs (or to cut taxes if that's your thing). Likewise, using AI to detect illegal sexist or racist material, or to restrict fake news, is a good thing if it is applied in an appropriate and proportionate way.

I also think that using AI tools to make my life just a little bit easier and less demanding is great. For example, using them to determine all the products and services that I have no interest in buying whatsoever. This is because it means I don't get bombarded with useless and annoying adverts. Instead, I'm only targeted with ads for things I might actually buy. Anyone born before about 1990 probably remembers the daily torrent of junk mail that used to pop through their letterbox, but these days it's very much a trickle, thanks in part to the use of AI technologies[104]. Likewise, all those cool apps and gadgets that I can buy to control my home, provide recommendations, organize my diary and so on, are excellent as long as the companies supplying these things don't abuse the data that they gather about me.

Things are also rarely clear cut. The underlying technology isn't intrinsically good or bad. It's that old adage about how it's used that determines if good or evil is dispensed. With the credit scoring example, the same type of artificial intelligence can also be used to identify those customers who are likely to struggle with debt repayments, to prevent them from borrowing more than they can afford[105]. Likewise, it can also be used to recommend more suitable, more affordable, credit products.

A similar argument applies to government surveillance tools such as facial recognition. Used proportionally, and in the right way, it can be a powerful tool to fight crime, find missing persons, speed you through customs when you go on holiday and a whole host of other things that can benefit individuals and wider society. A key feature with all these examples is one of intent. Are decisions being made for organizational gain or for the benefit of individuals or society?

Whether the intent is for good or ill, it's when AI tech is combined with huge amounts of personal data, that some of the most significant issues arise – what has been termed: "Surveillance capitalism[106]." The computers, when trained with vast amounts of very detailed personal information, can anticipate our intentions and future actions very accurately.

One way these insights are used is to reinforce our existing intentions, to make sure we carry through. If there is a tendency for us to do something that an organization wants us to do, then they will undertake activities to encourage us to do that. If we are talking about marketing, then it's all about exposing us to influences, such as ads and product placement, to tip us over the edge and buy something that we are interested in. Conversely, if our intentions are counter to what they want, then organizations will act to discourage us, to put us off doing that thing that we would otherwise have done. In this way, we are manipulated into taking a different decision to the one we otherwise would have. Negative campaigning in politics is a prime example. Imagine I'm a right (or left) wing politician. What I want to do is find the left (or right) wing voters who would normally be inclined to vote for a certain candidate, but who can be persuaded to switch their allegiance if I stream negative media stories about that candidate to them.

The other way organizations apply their AI-driven insights about us is to prevent us doing things or to force us down a particular route. In government, decisions about state benefits, identifying fraudsters, deciding which prisoners to grant parole too and so on, are increasing being made by AI-driven automated decision-making systems. If the computer says no, then that's that.

To their merit, these types of automated decision-making can drive efficiencies in government departments if deployed appropriately. This is by automatically determining the best options for benefit claimants, hospital patients, tax payers or whoever without needing to pay salaried employees to do it. However, badly designed systems can disadvantage the poor and vulnerable by leaving no route against which to appeal the decisions made by the machines[107]. Unless these automated decision-making systems are understood and sufficiently transparent, then the risk of harm, intended or otherwise, is very real. It's therefore very important to understand what decisions artificial intelligence is being used to make about us, what data is being used and what impact these decisions have upon our rights and freedoms. Perhaps most importantly, we need to understand who designs and controls these systems and that we understand their goals and motivations.

One of the biggest, high profile, concerns when it comes to the use of AI-driven decision-making is the issue of unjust discrimination; by which I mean *unfairly* favoring people with one type characteristic over those without it. Let's make no mistake, pretty much all decision-making systems, AI-based or not, work on the principle of discrimination. If we go back to the scorecard of Figure 4, we can see that in getting a good credit score, the scorecard favors those who have high incomes, own their own homes and so on. You are disadvantaged when it comes to getting a loan if you rent or don't earn a lot, for example. If you don't have discrimination then you don't have informed decision-making. You might as well follow the Dice Man's[108] example and make all your important life decisions by just rolling the die. Another take on the issue of discrimination is not if discrimination exists, but is that discrimination something that society deems to be unfair (unethical) or illegal? i.e. biased.

It's a mistake to think that AI algorithms naturally create bias in decision-making. Rather, what they actually do, is highlight (and possibly amplify) the biases that existed historically. They hold up a mirror to our own human failings when we've introduced bias when deciding how we treat each other. The problem really starts with the training data that we've created, not the algorithms that learn from it. It's a little sad, but very educational, to learn that as artificial intelligence becomes more widely used, so we are seeing just how biased human beings have been in the past, in almost every walk of life, to some degree or other.

Although we tend to talk about bias as a single thing, there are actually two different manifestations of bias found in the world of AI. The first type is historic **decision bias**. If we continue to think about the scorecard of Figure 4, then the reason eye color features in the scorecard is because in the past, lenders unfairly discriminated against non-white loan applicants who were more likely to have certain eye colors. When two people with identical profiles applied for credit, white people predominately got credit while non-white people had a greater tendency to be declined. The training data captures the flawed lending decisions made in the past, and hence, these flaws are carried forward into the scorecard that was developed using that data.

Another well-known example of racist decision bias is in parole granting decisions in the US. For several years, algorithms have been widely used in the decisions about which prisoners should be granted parole on the basis of a "Likelihood to reoffend score." Get a low score and you get your parole, but score high and you stay in jail. It has been suggested that some of these algorithms are fundamentally racist[109]. This is because they have been developed using historic data that was itself racist in nature. The algorithms didn't decide to be racist. They just turned out of that way because the training data was imperfect.

The second type of bias is *sample bias*. In this case, there is nothing wrong with individual items of data or the decisions that the data captures. Instead, certain groups are under or over-represented. The training data may contain disproportionately too many young people, under-represent people with disabilities, have more people from one gender than another and so on.

If sample bias exists, then ideally, one would try to obtain more data for the under-represented groups. However, that's often not possible. Therefore, one solution is to apply *weighting*, to rebalance the data that does exists. The data is manipulated to produce a better, less biased and more representative sample. If the training data contains predominately information about young people, then data about older people can be forced to be given more emphasis by the training algorithm[110]. Another solution is to exclude some of the data about younger people, which would also act to even things up[111].

A well-publicized example of this type of bias was seen with the role out of facial recognition software. Many of the initial facial recognition systems that were developed were created using training data that contained pictures of predominately white men. This data may have been gathered perfectly legally, with the privacy and data protection rights of those individuals being managed ethically and within the law. However, it should have been no surprise that when police forces started to use these systems to spot suspected criminals in public places, women and people from non-white ethnic backgrounds were subject to more errors and mistakes than white men[112]. This was because the data didn't include a diverse enough range of people to ensure that there was a consistent level of accuracy for everyone. In practice, this meant that they were stopped, searched and arrested more often than they should have been. Some organizations have rolled out face-recognition systems even when they knew in advance that these types of problem existed[113].

Bias is one issue that we need to understand and be aware of when talking about artificial intelligence based decision-making, but another very important issue is social exclusion. In many walks of life, there are large groups of people that have become increasingly sidelined from mainstream society. Artificial intelligence isn't solely responsible for this, but it may be a contributing factor.

We may all be individually assessed by an unbiased AI-based decision-making system, with everyone having their own unique score or prediction made about their behaviour. This in turn leads to a certain decision being made about how to treat them. Therefore, if the system is unbiased, in one sense, everyone is considered fairly on their own merits. However, people with similar, "Negative", traits will always tend to end up on the wrong side of the decision-making process. If you get offered a raw deal for one type of product or service, then its highly likely you'll get a raw deal for other things as well. The people who can't get a mortgage also tend to be those that can't get insurance, a secure job and so on.

One reason for this is the highly accurate, real-time, predictions that artificial intelligence enables to be made about people. As AI-based decision-making becomes better, so everything becomes more precise and clear cut. In some areas this is great. If I'm worried I've got cancer and an AI-based diagnostic tool can determine if I have the disease with 99.9% accuracy then that's great. Particularly if it can do a better job than a human doctor. But from a societal perspective, improved accuracy is not always a good thing.

If we think about insurance, for your car, house, healthcare or whatever, then the whole thing works on the premise of pooled risk. Everyone pays their premium into a central pot. If someone has a mishap the pot pays out. Over time, insurers have come to realize that some people carry a higher risk than others. Therefore, some people pay more for their insurance. I think that that's fine to a degree, but let's imagine that the insurer can make near perfect predictions about people. Let's say that an insurer predicts that I have almost zero chance of making a claim, but estimate that your chance of claiming is pretty much 100% What happens then?

In this scenario, having insurance becomes pointless for everyone except the insurer. This is because the insurance company will only provide insurance to people who will almost never claim and hence don't need insurance (me in this example). Those people that need insurance can't get it (that's you!) An extreme example maybe, but this is the direction of travel as more data about us becomes available and the AI-based decision-making tools that organizations use become increasingly accurate in predicting our behaviors[114].

Now, some would argue that these actions are fine – they are examples of businesses becoming ever more efficient for the benefit of their shareholders, which is what most companies are obliged to do under their articles of association[115]. Others however, would argue that it creates a divide in our society. Some, perhaps most, benefit from the advances that artificial intelligence brings in terms of better products and services, targeted pricing and so on, but a growing minority are being marginalized to a greater degree than we have seen in previous generations.

Differential pricing for risk-based products such as credit or insurance can be argued for on the basis that higher prices are needed to offset the increased risk of loan default and insurance claims respectively. However, differential pricing can also be applied to normal goods and services such as clothing and groceries. Retailers can determine who is likely to pay more for certain goods than others. If I'm a time poor individual with a considerable disposable income, then a retailer can use AI to identify me, and then advertise things to me at a higher price than my more discerning neighbor. I'll buy them because I can afford to and I don't have time to shop around. I put more value on convenience than getting the lowest possible price and for me that's fine. However, what can occur is that the poorest in society, who have less mobility and choice, end up paying the highest prices because they have fewer options as to where they can buy their goods and services from[116].

Consumers are very wary of differential pricing and generally view it as an unfair practice[117]. However, one way retailers have got around this problem is by approaching things in a different way. Headline prices are the same for everyone on their website, but everyone is open to receive a different discount coupon. Alternatively, if you hesitate before buying, by leaving goods in your basket, then that may prompt a discount that you would not have received if you bought immediately, without hesitation. Different means, same outcome. Another strategy is to have very similar (but subtly different) products priced differently. The retailer then directs customers to each product depending on what they know about each customer's buying behaviours.

Let's take another example, dating. These days, more Americans meet their partners for the first time online than in person[118]. Many of these first meetings occur via dating sites. How are you introduced to prospective partners? Most dating sites have details of thousands, if not millions, of people. You can't review everyone's profile, so the computers rank everyone in terms of their compatibility. The computer then shows you potential matches starting at the top of its list. Increasingly, it's complex algorithms that predict who is likely to be a good match with who, and it's the people behind the website who decide what constitutes a good match. It's them, not you, who ultimately decide what type of people you will meet. It might not be the case for every dating site, but most are commercial ventures. If they are ruthlessly following the profit motive, then it will be in their interest for you to continue to meet new people so that you use their service for as long as possible, rather than helping you find one long term partner – if that's what you want.

Then we come to the issue of diversity. Sure, you can choose from thousands of potential dates, but critically, there are whole groups of people, with certain tendencies, hobbies and behaviours, who you'll probably never be introduced too. This is simply because the algorithms deem that they won't be a good match for you. But what does that mean? What type of people are the algorithms keeping apart from each other and who is being forced together? Does this approach encourage people to mix between different social, ethnic and economic groups, or does it promote greater segregation by ensuring that you always meet people that are just like you? I'm afraid I don't know the answer to that because I don't have access to the relevant algorithms, but these are the types of question that we should be asking.

And it's not just dating where we see a potential loss of diversity. In theory, we now live in a world of almost endless choice in terms of the content that we can access via all of our various devices. However, the same types of algorithms used in dating are also used to present us with recommendations for all sorts of other products and services[119] This ranges from the books and music we read and listen to, to the news and websites we are directed to on social media to the products we buy. The whole thing acts as a feedback loop. The more we follow the recommendations we are presented with, so the more of those types of recommendations are suggested to us. If all you do is listen to music that the algorithms suggest, or only read posts that your social media platform serves up, then your whole outlook becomes increasingly narrow.

The fact that artificial intelligence is primarily being used to monitor and manipulate you, so that you do what governments and corporations want you to is more than a little depressing, but things are not all doom and gloom. As we mentioned at the start of the chapter, there are also lots of positive ways that AI is being applied to make our lives better. There are a lot of clever and principled people out there trying to use AI for positive purposes and for the benefits of society. It's also possible to make a lot of money and provide useful services at the same time – capitalism and social good are not always mutually exclusive concepts.

Resistance to the unfettered applications of artificial intelligence, and to how our personal data is used, also continues to grow. In the early 2000s, tech companies could do almost whatever they wanted to with your data. So, they used it to make as much money as they possibly could at your expense. However, the tide is now starting to turn. As we'll talk about in the next chapter, many governments are increasingly thinking about the ethics of personal data and automated decision-making (based on AI technologies). Consequently, many are reacting, albeit somewhat slowly, to address the concerns that have been raised about who has access to our data and how that data is being used.

8. Ethics and the Law

"A robot may not injure a human being or, through inaction, allow a human being to come to harm."[120]

Does a robot have human rights? If a computer creates a piece of art, does the computer own the copyright to it, the writer of the computer program or the artists, whose images were used to train it? If a self-driving car has to choose between crashing into three old people or two children, which is the best option? Should we put badly behaved robots in jail?

Whoa. Hold up there cowboy! I think we are getting a little ahead of ourselves. These are all very interesting and valid questions when having a theoretical discussion about the long term future of artificial intelligence. But how relevant are they today or even in the next decade? Are they just interesting hypothetical chit-chat about the distant future or practically important ethical concerns that we need to be dealing with right now?

It's fair to say, that it's these types of questions that get most of the media attention, and let's face it, they make for the most interesting discussions around the dinner table (well, maybe if you like that sort of thing). So yes, by all means let's have a good old chin wag about them. However, they can also be something of a distraction when it comes to more immediate and important questions about the ethical impacts of artificial intelligence on our lives today.

I mean, when was the last time you heard about someone having to make a split-second call about which group of pedestrians they were going to smash into when driving? Plausible, yes. Probable no. So, why make a big deal about it when talking about autonomous vehicles?

It may be theoretically relevant for a self-driving car as to what decision is made when it finds itself in this type of no-win situation, but it's a very trivial thing in the global scheme of things. Arguably, when it comes to self-driving cars, the much more important questions to be asking are about overall casualty rates of human driven cars versus those of autonomous vehicles, or broader questions about peoples' rights to drive versus fewer accidents if only self-driving cars are allowed on our streets.

Getting hung up on theoretical situations that could feasibly happen, but are very unlikely to do so, is an easy trap to fall into. It's also easy to spend all your time arguing about the moral rights and wrongs of certain situations that may seem vitally important, but which would only have very minor impacts if they occurred. At the same time, it's possible to ignore much more mundane moral decisions; decisions that are being made all the time in the real world, that are having very significant impacts on our rights and freedoms right now today, not at some distant point in one possible future.

I've got no problem with planning ahead and trying to address the problems that yet to be invented super AI may create, but let's not do it at the expense of the very real issues that are facing us in the here and now. In fact, if we don't pay enough attention to today's problems with artificial intelligence, then that could be a first step towards many of those possible future problems manifesting themselves sometime down the line. If we get a tight rein on AI now, then we can evolve and develop that as we need too, rather than facing a crisis at some point in the future.

As we discussed in the previous chapter, in the world today, the major concerns about the uses of artificial intelligence relate to how organizations are using it to influence and control us. Some of these activities may be considered bad, or even downright evil, but many, and possibly most, are perfectly good in their intent and lead to better outcomes for many.

What's difficult is that things are often not clear cut. AI is a just a tool in the same way a knife or a pen is a tool, albeit a very powerful one. Just like a knife, it's the user of the AI-based tool whose morality we have to question, not the AI app itself. Maybe, if artificial intelligence becomes sentient at some point in the future, then we'll need to address the morality of the actions taken by an AI that operates as an independent actor, but not at the moment. All current AI is directed and controlled by people in one way or another.

Therefore, when discussing the ethics of artificial intelligence, this is the area we are going to focus on. We are not going to get into a discussion about robot rights, or who is liable if a robotic surgeon bungles your operation and such like. These are all sensible topics but we'll leave them for another day (or maybe another decade). Instead, what we are going to look at are some of the ethical considerations about how governments, corporations and other entities are using AI to interact with and control people. We'll also discuss some of the legal frameworks that can and are being deployed to manage and guide those who develop and control artificial intelligence technologies.

In the previous chapter we raised the issue of unfair discrimination (bias) that can occur when AI-based decision-making systems are used. From an ethical perspective, what constitutes unfair discrimination is not as straightforward a question as you might think. For example, is it ethical to use someone's gender when making decisions about them (using AI or any other means)? Is it justifiable to treat men and women differently just because they are a man or woman? What about their age, sexual orientation or ethnic origin?

Some other questions to ponder: If an AI-based decision-making system displays some degree of bias, but less bias than when people make the decisions is that a step forward? Must we demand a higher degree of ethical conduct from a machine than a person? If less than perfect is acceptable, then what is the threshold of acceptability to set?

When it comes to life or death issues, like self-driving cars and medical robots, perfection seems to carry a lot of weight. We expect these systems to be near perfect rather than just being equal or better than us. A single fatal crash involving a self-driving vehicle, or a botched medical procedure by a robot, is enough to put things on hold. In the meantime, dozens are dying every day from accidents caused by human error. Should all AI-based systems be judged to the same high standards or can we be more lax with some types of artificial intelligence than others?

So how do we go about deciding what constitutes ethical use of AI-based technology when it comes to deciding how people are dealt with?

Before discussing the ethics of artificial intelligence further, I think we should be clear as to what we understand ethics (morals) to be. It's very easy to just jump in and start a discussion about the ethical aspects of something but forget that ethics isn't singular, objective or absolute. There are a lot of different opinions and a lot of grey areas to contend with. So, on that basis, we are going to go on a bit of a diversion for the next few pages and talk about ethics as a general field of study, before returning to the more specific topic of the ethics of artificial intelligence.

I have seen some very complex definitions of what ethics is, that I imagine would confuse even a PhD philosopher, but a very simple working definition, that seems to work pretty well in most situations, is that:

- Ethics is about right and wrong.

- It's about how we should treat each other and how we should behave.

What's not so simple is that ethics is personal and subjective. By that,

I mean that your ethics is different from my ethics. No two people have exactly the same ethical perspectives on everything and there will always be cases where people disagree. We all have a different world view. Sure, we may agree on many things, but we won't agree on everything. We may both vote Democrat and believe in social justice, but I think it's irresponsible to have an automatic assault rifle in my home for self-defense because it could fall into the wrong hands. You, on the other hand, argue that it's every citizen's duty to contribute to the safety of the neighborhood by being suitably prepared to defend it by whatever means necessary.

Having said that, there are some common ethical frameworks that can describe many of the approaches people take to deciding what's the right thing to do in a given situation. These frameworks are useful because they can be applied to all sorts of moral questions, not just those about the use of artificial intelligence. They can also help to explain why someone may disagree with you when you put forward what you think is a very simple and clear-cut case. It's not that they are evil, or are just arguing for the sake of it, they just happen to have a different approach to assessing what being ethical means than you do.

A key question when it comes to forming an ethical stance on something is if you are coming at things from a consequentialist or non-consequentialist perspective. Consider Figure 15.

If you tend to judge things in terms of the end game, where anything goes if you get the right outcome, then that's very much a consequentialist view. The ends justify the means. However, if you believe that it's important to follow a certain moral code in getting to your goal, then that's more of a non-consequentialist way of looking at things. For a non-consequentialist, doing things in the right way and following your principles is what matters. This is even if it means that the final outcome isn't as good as it could be because you constrained your actions in some way. It's just as much about your journey as it is about the final destination.

Figure 15. Ethical Frameworks

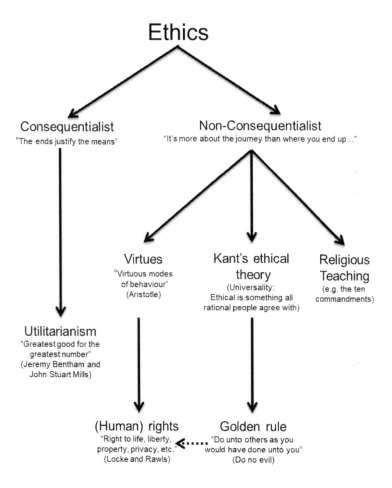

To illustrate these two perspectives consider the following statement:

> It is ethical to kill someone if it saves the lives of at least two other people.

Do you agree with this statement? If your answer is an unequivocal yes, it's always the right thing to do, then this consequentialist view could be taken to mean that you think it's OK to perform medical experiments on a group of people, which kills them, if it delivers a treatment that saves more lives in the long run.

However, If you disagree with the statement, on the principle that it contravenes basic human rights not to be forced to participate in such experiments, even if it means a lot of people are going to die from the disease, then this implies that you are adopting a more non-consequentialist perspective.

This is quite a simple and simplistic example. I can present other arguments that might persuade you to think differently. I could argue, for example, that killing people in medical experiments causes significant harm to wider society. This is due to the fear and bad feeling created by the chance of becoming a test subject and from the emotional damage inflicted upon the relatives and friends of the test subjects.

In this case, the consequentialist may decide that the action is unethical based on some more general measure of wider public good, rather than just considering the consequences for the individual test subjects. From a non-consequentialist viewpoint, it can also be argued that the *principle* of saving many lives is right, and therefore, killing the test subjects is ethical.

So, hopefully, this theoretical example demonstrates that the ethical rightness or wrongness of some issues is not always clear cut. The context of the problem and the scope of application both have a part to play.

There are lots of different forms of consequentialism, but the best known one is Utilitarianism. If you are a utilitarian, then you tend to think of an ethical action as being one that leads to: "The greatest good for the greatest number."[121] An ethical action is one that maximizes happiness and pleasure within the population. For a utilitarian, an action is only unethical if it leads to the opposite; i.e. pain or unhappiness.

For utilitarianism to be practical in the real world, you've got to be able to measure the amounts of happiness and pleasure that different actions cause when you make a decision to do something. This is so that the benefits of different decisions can be compared and the best one chosen.

What this often means in practice is that everything gets assigned a monetary value, a cash amount, as a measure of its desirability. The outcome that yields the greatest overall value is taken to be the most ethical. You can see utilitarianism in action whenever politicians talk about the cost/benefit case for where to build a new hospital, airport or whatever. All the upsides and downsides for building it in each location are assigned a dollar value. Everything is then totted up to give a single value that represents the overall net benefit of each option. The option with the greatest net benefit is chosen.

That's enough consequentialism for now, but we'll come back to it later. Let's flip things over to the other side and look at some non-consequentialist models of ethical behaviour that people can adopt.

Religion provides one example of a non-consequentialist approach to ethics which forms the basis of many people's lives. God-given laws define the ethical framework that they operate within, on the basis that God is undeniably a better judge of right and wrong than we are. The Ten Commandments, given by God to Moses, are possibly the best-known examples of these. "Honor your father and mother", "Do not steal", "Do not covet your neighbors wife", "Do not kill" and so on.

However, no system of religious laws deals with everything that people encounter in their daily lives. Reason is often required to interpret and apply what is believed to be God's will in specific situations. There are no holy scriptures about climate change or the use of chemical weapons (or the use of AI), but it is still possible to formulate a theologically derived moral stance on these issues.

Another non-consequentialist view of ethics was put forward by the German philosopher Immanuel Kant. What Kant proposed was a model of ethical behaviour based on duty and a respect for others. Instead of asking what the impact of an action will be on someone, Kant argued that an action is ethical if it is shows respect for other people. To put it another way, people should be encouraged to act in good faith, out of a sense of duty as to what is right and proper.

The fundamental concept captured in Kant's thinking is the concept of universality[122]. Ethical behavior is something that all rational people agree with. Imagine that you and several of your colleagues are working late in the office one night. To keep your energy levels up, your boss orders in a nice big deep pan pizza for you all to share. Now, everyone might want the entire pizza for themselves, but the only sensible (and hence ethical) decision that any group of rational people could arrive at is to share the pizza equally amongst everyone.

Simple hey? – you don't disagree with that do you? No? OK, then why don't we apply the same principle to society at large? Why don't we pool all of our assets and share them equally with everyone? If you are a communist then I guess you might agree with this idea, but I suspect that most of you won't. Again, the context of the problem is just as important as the principle.

Kant also presented another set of ideas, complementary to the principle of universality. He proposed that people should never be objectified. Everyone should be treated as a sentient being with the right to their own life and opinions. People aren't just a natural resource to use and abuse. You shouldn't employ people as means to your own ends without consideration of the impacts that your actions may have upon them.

Kant's thinking is often expressed as the "Golden Rule." Which states:

"Do to others as you would have done to you."

The problem with the golden rule is that it implies that we should do stuff like giving all our worldly goods to random strangers. Why? Because this is what we'd want them to do for us. Perhaps it's not surprising that it's the inverse of the Golden Rule that is most practical in the real world: "Do not do to others as we would not have done to us." To put it another way, don't do things that harm others or even more simply: "Do no evil."

What a great idea for a company slogan, and for a while it seemed that that was what Google was going for as it's unofficial motto[123], but sometime in 2018 they decided to remove it from their mission statements. Shame.

The Greek Philosopher Aristotle believed that for everything in nature there was a right and proper purpose – a natural law to which it should conform. Aristotle understood human beings to be part of nature, and therefore, argued that this principle applied to people and their activities too. Our ultimate purpose is to be fulfilled in life and Aristotle believed that one way for us to achieve this was by undertaking social responsibilities. Examples of this are being a good parent or a conscientious worker. However, fulfillment also means acting as a rational being, guided by reason. This is so that we control our emotions, live disciplined lives and don't give into sudden temptation to do stupid or outlandish things.

Aristotle termed these principles 'virtues' and he concluded that there were inherently virtuous modes of behavior. Over time, these ideas evolved into the concept of natural rights (human rights) as we know them today[124]. There are certain things that an individual has a given right to do and certain things an individual cannot be denied. These rights are beyond the power of anyone, including governments, states and corporations to grant or deny.

In many ways, the concept of human rights can be considered a natural complement to Kant's ideas about duty and respect for individuals. So, for instance, I could argue that all workers have a basic right to receive a living wage for their labor. Meanwhile, employers have a duty to provide their employees with a good working environment and fair wages[125].

Ethical theories are all very well but, in their pure form, all ethical frameworks can be found wanting when applied to many real-world situations.

During the twentieth century, Soviet Russia applied what can be described as extreme utilitarian ideals to government policy, putting the welfare of the state before that of the individual citizens that comprised the state. Millions of people were deliberately killed as a result of various state policies to control any dissenting voices that challenged the state's view as to what was in the best interests of the wider population.

Religious teachings have often been distorted to provide justification for wars, much cruelty and many acts of terrorism throughout history. It can also be argued that to follow a Kantian or human rights-based philosophy can lead to equally poor decision-making. In the extreme, following an "Always do the right thing" approach can conceivably lead to terrible disaster because someone does what they believe should be done, not what avoids some ultimate catastrophe.

We see this played out in Hollywood disaster movies all the time. The hero has to take a diversion to save a small child, their lover, family member or even their pets while the clock is ticking to save the entire world. Rationally, they should just get on with the world saving, but in the eyes of Hollywood, the hero has to do what is right. They can't be seen to leave someone behind, even if in a real-world situation that would clearly be the best thing to do.

One real-world example of this type of thinking, to justify denying people their basic human rights for the greater good, comes from the early 2000s when it was used by the US and UK governments to justify the way 'suspected terrorists' were treated at Guantanamo Bay and Belmarsh Prison respectively.

The argument that these two governments put forward was something along the lines of:

> "We need to do horrible things that are against the basic principles of our society, such as waterboarding prisoners, to get information. This is because if we don't do everything possible to get the information, then there is an increased risk of more terrorist's attacks."

On this basis, maybe I could argue that we should consider torturing everyone who drives a car badly. This is because it will act as a deterrent against poor driving and saves lives, given that every year far more people die in the UK and US in car accidents, caused by dangerous driving, than in terrorist incidents.

Ethical theory can be easy to apply in certain straightforward cases but many real-world situations are extremely complex. The full facts may be disputed, unknown or ambiguous. Peoples' and societies views can also shift and change, or even completely invert, over time. I'm sure you won't have to think very hard to come up with all sorts of activities and views that were considered immoral (and often illegal) in many regions for centuries, but which are now considered completely normal. If you expressed those same views today, you could well find yourself being prosecuted for them or, at the very least, being pilloried on social media.

It's also the case that different societies can have different value systems, particularly if they have different religious or cultural perspectives. People with a traditional Western (North American\Western European) heritage may consider it unethical (and illegal) to bribe a government official to speed up government bureaucracy. In a lot of countries however, paying a civil servant to move your case to the front of the queue is just normal day to day practice. Everyone does it, and even if it's technically illegal, the authorities will always turn a blind eye. In some cases, things won't get done at all unless a suitable gratuity is offered.

When people hold opposing views it can be very difficult to form a consensus as to whether a particular action is right or wrong, moral or immoral. There are lots of highly controversial issues, including animal rights, the use of nuclear weapons, abortion, euthanasia and capital punishment that all fall into this category. In each case, it's possible to argue different but convincing cases from more than one perspective.

In life, people naturally adopt ethical positions that encompass both consequentialist and non-consequentialist perspectives. They then come up with their own individual view as to what constitutes good behavior. What this means is that no two people will share exactly the same viewpoint on every subject. Everyone will disagree to some degree as to what is deemed to be morally the righty thing to do and what is not.

However, what can be said to apply across all ethical frameworks, is that:

> Ethics carries with it the idea of something more important than the individual.
>
> An ethical action is one that the perpetrator can defend in terms of more than self-interest.
>
> To act ethically one must, at the very least, consider the impact of one's actions on others[126].

On this basis, I'd argue that a company that blindly follows the principle of maximizing their profits or a government acting only to maintain itself in power are two clear examples of entities behaving unethically.

There you have it. A whistle stop tour of the fascinating topic of ethics. Let's now bring things back to the ethics of AI-driven tools, that are used to assess, influence, manipulate and control people. As I see it, there are four key aspects of the problem to consider.

1. **Data.** The personal data that organizations hold about us. So, that's all sorts of stuff from our income and age, to our browsing history, location data, medical records, Tweets, Facebook posts and so on.

2. **Purpose.** What reason does someone have for using our personal data to make decisions about us? Are they using it to target us with adverts, decline people for credit or insurance, identify someone as a potential criminal? or whatever.

3. **Intent (Beneficiary).** For whose benefit are the data and purpose being employed? Is it for the individual in question or someone else? Are you trying to help people or help yourself? For example, is your health app intended to make you money or reduce my chance of getting ill?

4. **Mechanism.** The way that data is used to achieve the desired purpose. One aspect of this is about who makes the decisions; i.e. is it a human being or an automated AI-driven decision-making system? The other aspect is an understanding of the underlying logic used to arrive at a decision, regardless of whether it is human or AI-based.

It may seem that artificial intelligence is only relevant to the final point, but the first three items about data, purpose and intent are just as important. This is because collectively, they drive how artificial intelligence applications are developed and deployed, and how their impacts are assessed in terms of benefits and drawbacks. Therefore, when discussing the ethics of artificial intelligence, we need to consider these four aspects collectively if we are going to present a rounded view.

Starting with data, how we deal with peoples' data is invariably linked to how artificial intelligence is used and how powerful it can become. This is because AI can only operate effectively when it has sufficient volumes of data to work with. This includes data resources to build intelligent tools in the first place but also the data required to deploy them. The credit scoring model in Figure 4 can't generate an accurate credit score for someone if you don't know what that person's income, eye color, employment status and so on are. We can exert a lot of control over how AI is developed and deployed by controlling the data that it needs to function. Restrict the data artificial intelligence has access to and you reduce its ability to anticipate your behaviours, and therefore, the degree to which it can control you.

In terms of purpose and intent, we have to remember that current AI only does exactly what it has been tasked to do. Mis-specify the task and it won't do what you really want. Therefore, purpose is vitally important. Not least, because the purpose is often driven by the intent of the developer. They want the AI to achieve something that they have defined. If the primary intent is to generate profits, then that may drive a very different view of purpose than if the intent is to save lives. This is even if the application is being described in exactly the same way in terms of what it is meant to do.

Profit objectives and individual benefit aren't mutually exclusive, but there is a pretty torrid history of large corporations chasing the profit motive at the expense of individuals. Examples range from the tobacco industry, to the VW emissions scandal, Cambridge Analytica and pharmaceutical cover-ups to climate change denial in the fossil fuel industry. The list goes on and on. Even if the intent is altruistic in nature, if the right purpose isn't used in training the model that drives an AI application, then that could lead to some very poor, if unintended, outcomes for people.

Finally there is the question around mechanism. Regardless of the data, purpose and intent, how are decisions about people being made. What checks and controls are in place to ensure that the decision-making mechanism is acting in a legal, fair and ethical way?

Personal data tends to get most of the press, so let's continue to think a bit more about personal data and data protection.

As described earlier in the book, machine learning based AI, which covers pretty much all practical AI systems in use today, inherently discriminates. That is how it works. It looks for differences between things to identify what features of the environment are correlated with different outcomes. So, what data is it OK to use about people? Let's consider Figure 16.

Figure 16. Whose data is it anyway?

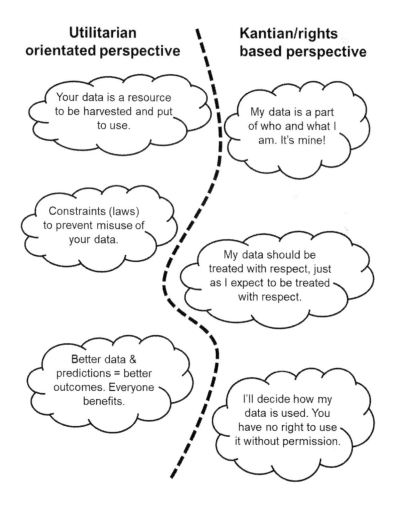

Earlier in the Chapter I asked if it was ethical to use gender, age, sexual orientation and ethnic origin in decision making. To be honest, this was something of a trick question as you'll get a different answer depending on what ethical perspective you have adopted and the use to which the data is being put.

From a consequentialist (utilitarian) perspective then yes, it's easy to argue that it's OK to use any or all of this type of data if, on balance, it results in good outcomes for people; i.e. better outcomes than if you didn't use this data.

From this viewpoint, data is a resource to be mined and harvested. If we can hoover up all that great data about people and use it for productive purposes, then that's fine, as long as overall we get some net benefit. Letting Google, Apple, Amazon etc., have access to everyone's data is a good thing if they use it to develop better diagnostic tools for doctors, improve educational outcomes or increase profits (and hence wealth) and so on. If the benefits justify it then it's a no brainer. If a few people are hurt or disadvantaged as a result, then that's a price worth paying. If, for any reason, using that data causes problems that we feel must be addressed, such as protecting people's medical data, then we'll put in some specific laws to deal with that particular problem.

From a rights-based perspective things look very different. The starting point is very much that my data is mine and your data is yours. We own it. It's a violation of our rights for someone to gather or use that data without our permission. This is regardless of whether or not you or they will benefit (or be disadvantaged) from using it. In principle, it's no different to a violation of our person or our property. If someone "Borrowed" your car without asking you first, because it meant that they got some benefit from using it, that's not right, even if it didn't disadvantage you in any way.

A classic issue in this space is the use of our medical data. Applying AI to vast quantities of medical data can, without doubt, bring many advances in the understanding, diagnosis and treatment of many diseases. However, health authorities and government regulators are very concerned about the rights of individuals to decide how their sensitive medical data is used and this has restricted how much access tech companies have had to this type of data to date.

These differences, between utilitarian and rights-based approaches, are clearly visible in the different paths that US and EU authorities have followed when it comes to Data Protection law. In the US, things have been driven very much from the utilitarian camp in a piecemeal way. When there is sufficient concern over a specific way in which data is being used, then a law is passed to address that concern. This principle dates back to the 1970s, with one of the first US data laws being the 1974 Equal Credit Opportunity Act (ECOA)[127] The ECOA refers to the type of data that can be used in credit granting decisions. Likewise, when concerns were raised over how private medical providers were using their patients' data, The Health Insurance Portability and Accountability Act 1996 (HIPAA) was enacted to address these concerns. There are a host of other similar laws at both state and federal level dealing with different data issues.

In Europe, the approach has been very different. Since introducing the first international data protection law back in 1981[128], and more recently with the General Data Protection Regulation (GDPR) in 2018, it's been all about individual rights. In particular, the GDPR provides individuals with comprehensive rights over how their data is gathered and stored, and the uses to which it is put. For example, using data about peoples' race, political views and medical history is explicitly forbidden in the GDPR, unless certain specific conditions have been met[129].

The GDPR is also heavily principle based. It's very much about the spirit of the law and doing the right thing, just as much as it's about compliance with the letter of the law. Consequently, very little in the GDPR is industry specific, it's very much a general approach that is intended to operate across every government and industry sector.

Being principle based also means that in most cases the GDPR doesn't tell organizations exactly what they must do to comply with it. Some argue that this puts a large and expensive burden on organizations as they figure out what they need to do, but the idea is to get them to think about their responsibilities to the people they hold data about. It's meant to encourage them to act in a responsible way, rather than blindly following a rigid set of regulations.

These two different perspectives are one reason why US based companies such as Amazon, Google, Facebook and others have, in the past, struggled to find common ground with EU regulators over how personal data can be gathered and used in European countries, and hence, why it has been EU regulators that have imposed the some of the biggest fines ever seen (some running to hundreds of millions of dollars) on companies for various transgressions of data privacy laws.

The GDPR isn't perfect – no law ever is. To remain current, it will need on-going updates to deal with new ways in which data is being used (such as facial recognition, which was not well covered in the original legislation). However, the GDPR represents a significant step forward in providing people with real protections about how their data is gathered and used. As at today, it is widely accepted that it sets the gold standard for data protection legislation anywhere in the world.

Looking forward more widely, the general trend appears to be towards the EU view, with state and federal legislation in the US increasingly incorporating greater elements of the data rights of people[130]. However, there is still a long way to go before US data protection laws provide the same protections as those found in the EU. But let's not forget, that in some countries, data protection laws are non-existent. Governments and other organizations can do whatever they want. So, maybe an emphasis on those regions where individuals have very few rights regarding their data is where our attention should be focused first.

We could talk about personal data and data protection all day, but let's now move on to think about some of the issues that arise when organizations use all that data to decide how people are treated using AI-based decision-making tools; i.e. item 4 in my aforementioned list. In talking about decision-making, two (linked) subjects that often come to the fore are those of bias and explicability.

When it comes to bias, at one time, it was widely believed (somewhat naively) that because training algorithms could find the optimal way to interpret the training data, that meant that the resulting model could not show any bias. In one sense that's true. The training algorithm used to build a model doesn't start with any preconceptions and doesn't display any conscious or unconscious biases. It just blindly considers the data it has been presented with and learns from that. Unfortunately, what is increasingly evident, is that our past human-made decisions were often sub-optimal and biased in all sorts of ways. Consequently, the training data used to construct the models used in many AI-based systems is fundamentally flawed.

In my experience, far too much emphasis has been put on trying to identify biases with AI-based decision-making tools *after* the event; i.e. seeing what biases exist in the model that has been created. To some extent, by the time the training algorithm has been applied and a model created, the horse has already bolted. Yes, we do need to be checking for bias in the final product, but the first priority should be given to examining the training data used to build AI applications before anyone starts to think about algorithms. This is to ensure that the training data provides a representative, fair and balanced view of the section of society that the model is going to be applied too. If the training data is unbiased, then it's highly likely that any AI application developed using that data will be unbiased to, or at least that the bias will be reduced considerably.

If we think back to the scorecard in Figure 4, then the most obvious concern that exists is the use of eye color. This is because it acts as a proxy for ethnicity. The data scientist should ideally have flagged it as a concern at the outset, and serious consideration given to excluding it from the training data used to create the scorecard.

We obviously need to be careful about using certain types of sensitive data about people, including their race, gender, religion, sexual orientation and so on, but that's not to say this type of data should not be used in some situations. There is certainly a case that if the purpose is altruistic in intent, undertaken for the benefit of the individual in question and without detriment to others, then allowing decision-making processes to use that data is ethical in some circumstances.

Certainly, in medicine, things like gender and race are correlated with certain medical conditions. On that basis, it would seem sensible to encourage people to allow the use this type of data to help diagnose them. Another example of altruistic purpose is identifying people who don't claim all the state aid/benefits they are entitled too. This is so that you can contact them and tell them about the extra help they can get. However, using sensitive personal data to set insurance premiums, where the primary beneficiary is the insurer, or to exclude people from receiving benefits, is a more questionable activity.

To date, bias is something that the GDPR, and other data protection laws have touched upon, but not directly or comprehensively addressed[131]. This is because they have tended to focus on getting things right at a personal level not a societal one. What I mean by this is that a law such as the GDPR forces organizations to ensure that they get your permission to gather and use your data, and that the data they hold about you is accurate and they use it in a fair and transparent way. However, at an aggregate level, there is nothing to say, that collectively, the data they hold and use should be representative of society and its values, or that when data is skewed, corrective action must be taken. Maybe the law should be amended so that any training data should be required to pass certain quality standards before it is allowed to be used to build AI applications. It should be required to obtain a stamp of "Unbiased assurance" before it can be used.

To be able to understand and deal with bias, data is key, but we also need to be able to pick apart how decisions are arrived at using that data. We need to understand the mechanism for decision-making so that we can in turn identify where and how bias is occurring.

If we talk to a human expert about how they come to a decision, then they can usually say well, we think this patient has cancer because of this dark patch on the scan, or we are going to arrest this suspect because we've got this piece of evidence that suggests we should haul them in for questioning. However, when it comes to algorithms things are often less clear.

Many algorithms, and in particular ones based on neural networks, are increasingly *black box* in nature. Black box means that you can determine what data provides the inputs to an algorithm and what outputs (decisions) come out, but you can't easily work out the way all that data was combined together to deliver one type of decision and not another. You can see this just by looking at the difference between the scorecard and neural network versions of the credit scoring models we've considered in previous chapters (Figures 4 and 7 & 8). With the scorecard, you can easily see what contributes to the final score and by how much, but for the network it's not clear at all.

There is a lot of research effort going into understanding how (deep) neural networks generate their outputs from a given set of inputs, but it is still very much an evolving element of AI-based-decision making. A lot of progress has been made, but there is still more to do before we have a set of standardized tools that automatically explain how AI generates the outcomes and decisions that it does.

Understanding why a model produces certain outcomes doesn't solve the bias problem, but it certainly helps understand what features contribute to the decision in what way. The GDPR also helps out EU citizens, here. In particular, it requires three key things to be done when fully automated decision-making is undertaken.

- Decisions about people should be made in a fair and transparent way[132].

- If the decision is an important one, that will have a significant impact on someone, then they have the right to have that decision put aside, and an independent decision made by a real human being[133].

- People have a right to an explanation as to how a decision was made about them[134].

How organizations should apply these principles is still evolving, but in the UK, the regulator (The Information Commissioner) has drawn up a set of detailed guidelines on explicability. These indicate that a very high degree of explanation will need to be provided in terms that normal people can comprehend[135]. If someone is using a huge neural network to decide if you can get a mortgages or not, then they will need to be able to dig into that neural network and unpick how it's making its decisions. They must then explain that to you in a way that you can understand. This isn't impossible, but it does require cost and effort. Consequently, organizations may need to reassess their cost/benefit case for using advanced AI models such as deep neural networks, compared to simpler, more explicable, models such as scorecards and decision trees.

The UK Information Commissioner has also made it clear that if an organization doesn't have the time or money to build explicability into its product development cycle, then it should seriously consider if it should be using that technology at all. Otherwise, if it fails to provide adequate explanations to its customers, it could face fines of up to four percent of its global turnover.

Even though the GDPR, and similar laws elsewhere, represent a major step forward in the fight to protect our data and how organizations use it, what is worth noting is that although such laws can be used to constrain and control organizational entities, they don't act to change their nature. All laws are imperfect, with gaps and loopholes that the unscrupulous look to exploit. Therefore, we can't rely on the law alone to force corporations to treat individuals in a morally appropriate way. For an organization to be ethical, it needs to include a degree of moral self-assessment of its actions into its operational processes and procedures.

A few years ago, I used to talk about the wild west days of artificial intelligence, but things are now moving out of that sphere, at least in the EU and US. If an organization wants to use AI-based decision-making tools then it needs to be able to demonstrate that it is acting in the interest of its customers as well as simply following legal compliance.

Explicability is central to us being able to trust AI-based decision-making. It also supports the identification of unfair discrimination and bias in the way automated AI tools treat people. By laying the groundwork now, to enforce developers to explain how their AI tools work, provides a key foundation for protecting us in the future as artificial intelligence becomes ever more advanced and ever more integrated into our daily lives. Isaac Asimov may have famously proposed his "Three laws of robotics"[136] that were designed to protect us against robot takeover, but if I were to define three laws for artificial intelligence today, then I would have explicability of action as one of them; that is: "Why did you do that robot?"

One argument against organizations putting more effort into oversight of their artificial intelligence tools and ethical decision-making is that all these new laws and consumer rights could have an impact on their costs and hence profitability. However, when it comes to corporate ethics, the evidence is that the opposite is true. Building ethical considerations into the fabric of corporate culture is actually a benefit. But how can that be? If you restrict yourself so that you don't always chase the dollar, then surely that means less profit at the end of the day?

There is a lot of evidence that if one is thinking about the long term, then adopting an ethical code of behavior delivers a real bottom line benefit.[137] One reason for this is that those organizations that try to "Do the right thing" reduce their risk of reputational damage. If the public decide that they don't like the way you operate, then that's the sort of tarnish that can take a long time to remove. It's several years since the Cambridge Analytica scandal hit the headlines, but it still comes up frequently when people talk about Facebook and data protection.

If we take all of the issues discussed in this chapter so far then, where organizations are investing in artificial intelligence technologies that significantly impact people, they need to ensure that they consider all of the following:

- **Ethical standards for AI use.** Organizations should have a stated set of principles about how they will use AI that they adhere too. For this to be meaningful, suitable audit and governance processes are required to manage and provide oversight of their decision-making systems, to ensure compliance.

- **Impact assessments.** There should always be an impact assessment performed whenever AI is being deployed. In particular, to identify any negative impacts on individuals or specific socio-economic groups and to identify any biases that may exist.

- **Independent validation and oversight.** External parties should be involved in auditing ethical compliance of artificial intelligence technologies. Companies can only do so much to self-assess without having independent oversight of their activities.

- **Human check points.** Organizations should not expect to do away with all human expertise entirely. There must always be a way to bypass the machines if there is some flaw or concern over a decision that they have made.

In Europe, principles similar to these are being recommended by regulators to ensure that organizations comply with the principles of the GDPR. In response, we are seeing an increasing number of organizations that operate in Europe adopting them as an integral component of their product development cycle when incorporating AI tech into their products and services.

We are also seeing some moves in North America, Australia and many other countries to include these types of controls as well. There is therefore, grounds for optimism. My hope is that we'll continue to witness the increasing involvement of governments around the world to ensure that where the data rights of citizens need to be protected and AI technologies controlled, regulators and law enforcers are sufficiently empowered to do so.

9. What Does the Future Hold?

"The future has already arrived — it's just not evenly distributed yet."[138]

I promised at the beginning that this wasn't going to be a book about some imaginary sci-fi future underpinned by artificial super intelligences, killer androids and the like. Instead, the purpose was to explain how artificial intelligence works and is being deployed in the world today, and the impact it is having across society. However, given how fast artificial intelligence technologies are being rolled out, I think it's worth spending just a little bit of time thinking about how things may develop in the next few years.

If I was a futurologist in the bright and optimistic days of the early 1960s, dreaming of a better world driven by the "White heat of technology[139]" then by all accounts everyone living in the twenty-first century would be enjoying a world with advanced health care, free energy and more leisure time than they know what to do with. Intelligent robots would be everywhere, helping and supporting us in our daily lives.

So why was utopia not achieved then and will artificial intelligence help us to deliver it now? "The tech is easy, it's the people that's the problem." That's not an atypical response from the Geeks, Nerds and PhDs. If we didn't have people to worry about we'd have had self-driving cars in every city of the world years ago. All of Africa's problems would have been solved by now, and we'd all be living in peace and harmony. But we do have people to worry about. Actually, without the people what's the point? Technology only has purpose when it's applied to what people do or how society behaves. Technology for technology's sake serves no useful purpose.

So, what does the future hold? If I was being philosophical and feeling a little pompous, I might be tempted to say: "Anything and everything. If you go far enough in time and space, then everything possible becomes reality. In an infinite universe all things that can happen will happen." That's fine, but in terms of practical real-world developments, a much better way to look at things is not if, but when and where things are likely to occur. What happens in China or California won't necessarily be what goes down in Germany, Nigeria or Venezuela. The technology we see in the smart cities of Silicon Valley next year, may not reach London, Delhi or Paris until years later.

We also seem to live in an increasingly divided world. Many basic technologies, that became readily accessible in countries like the US and the UK decades ago, have yet to reach much of the developing world. More than 100 years after the invention of the flush toilet, more than twenty-five percent of the world's population still don't have access to basic toilets or latrines[140]. Likewise, MRI and CAT scanners, and a lot of other health technology that entered everyday use back in the last century, are still a rarity in many regions. Looking ahead, I can't see any reason why things will be different this time around. What we are probably going to see is certain technologies move forward very quickly in some regions, but not universally in all regions and almost no progress made at all in others.

As I said back in Chapter 1, thinking that you know exactly what is going to happen in the future is a very dangerous (and possibly arrogant?) thing to claim. You also have to be very careful not to be taken in by all the interesting ideas that people think up and you hear about in the media, but never make it beyond the R&D stage. When I was a kid, there was a great British TV show called "Tomorrows World." Every week, the team demonstrated several new and interesting inventions. I thought the show was brilliant but

I remember it dawning on me that very few things ever seemed to make it into the shops or other aspects of real-life. Sure, there were a few things like CD players, touchscreens and internet banking that made it big[141], but the vast majority failed to get over the hurdles required to make them successful. I don't know precisely why all those inventions failed to make it but maybe the cost of putting them into production was prohibitively expensive. Maybe there were health and safety issues, or maybe it was a really cool thing to use once, but not the sort of thing that anyone would ever use practically or wanted to buy. However, at the time, nobody really knew which inventions were going to be a failure or a success. It was only years later that it could be determined which inventions had succeeded and which had not.

Consequently, so as not to risk making a fool of myself, I'm not going to make any specific predictions about this or that application of AI making it big in the next few years. However, barring the zombie apocalypse or complete climate collapse, then over the next five to ten years, the things I'd be willing to bet a few quid on are:

1. Existing artificial intelligence is only going to get better. It will become ever more prevalent in industry, our homes, schools, hospitals, and almost everywhere else.

2. There will be many more battles between technology companies and governments/regulators over the use of artificial intelligence and personal data. The initial skirmishes have already commenced.

3. We won't see the development of a self-aware, conscious Super AI with general intelligence, that is equal or better than our own, using current technologies (apology or sigh of relief as appropriate).

It can be easy to try and write-off some AI technologies due the flaws that they exhibited when they were first put into commercial use, but many applications of artificial intelligence are still at a relatively early stage of their development. They will improve and get better in time and existing weaknesses will be addressed. In chapter 2, I made quite a big deal about all those flawed object recognition systems, you remember, the ones that couldn't determine what an upside-down bus was and so on? Well, I don't expect it will be long before that problem is solved. In fact, I wouldn't bet against the possibility that it may already have been sorted out by the time you read this!

Other examples of problems with existing artificial intelligence applications, such as the gender bias that resulted in Amazon decommissioning its hiring AI[142], or the issues over racial bias seen in some facial recognition systems are mostly solvable problems. This is because the issues are mainly to do with the training data used to create the models, rather than the underlying technology itself. Get the training data right and the bias should disappear. As these problems are solved, we'll see ever more examples of this type of AI being deployed. Conceivably, in almost any area where decision-making capability is required and there is sufficient good quality data available to feed the training algorithms with.

The second area where I expect to see changes is in legislative and regulatory frameworks being developed to address some of the problems with personal data and its use within AI-based decision-making systems. When technology companies began using artificial intelligence at an industrial scale in mid-late 2000s, very few people really understood the scale of the impacts that their activities would have on individuals, the workplace and society at large. It was only as concerns over privacy, bias and social exclusion began to be raised that governments and regulators began to wake up to the risks that unrestricted access to personal data created. In particular, the danger of using that data to develop advanced AI-based technologies to manipulate and control people for commercial and/or political gain, without due consideration of the impacts on the individuals themselves. Restrict access to peoples' data, and that very much limits how much you can predict and influence their behaviour. Without data, you can't train the algorithms in the first place.

The adoption of social media, and other applications of new technology that thrive on personal data, came about very quickly. Governments and regulators on the other hand tend to move very slowly. As we've discussed, many tech companies adopted a "Move fast and break things" approach to advance quickly in the early 2000s, but regulators tend to come at things from the opposite end. "Ponder for quite a long time, consult widely and try and get it right." might be a more appropriate mantra for regulators and law makers. However, they are now getting up to speed. The EU in particular, is not afraid to challenge the tech giants and we are beginning to see very significant fines being levied for abuses of personal data in EU countries.

Prior to the role out of the GDPR in 2018, the largest fines regulators could levy in the EU for misuse of personal data was less than $1m[143]. Following the introduction of the GDPR, this is now up to four percent of a company's global annual turnover. The evidence is that the regulators are not afraid to exercise their powers, with several fines in excess of $100m issued in the first year after the GDPR was introduced.

EU member states are also continuing to evolve their guidelines on how companies should use and explain their use of AI-based technologies, when these are used to decide how to treat people. Various laws and regulations are being enacted in other countries, outside the EU, as well. Existing privacy and human rights laws are also being used to limit the powers that the tech companies have over us.

As a result of new laws and regulations, what I think will occur is a change in the skill sets that companies employ to develop AI-based systems. Historically, the goal for many organizations has been to recruit lots of very clever technical people with expertise in mathematics and computing (data scientists) who develop AI-based systems. In future, that emphasis is likely to change. This is because the biggest challenge for most organizations won't be the techy stuff. Rather, the biggest tasks will be implementing all of the constraints and controls required to ensure that organizations are operating within the law. In particular, increased staffing of compliance, governance and audit functions to oversee the end-to-end process for getting AI-based tools into production.

We can already see this in some areas, such as the capital models used for credit and insurance risk. The development of the base algorithms to predict credit and insurance losses are only a tiny fraction of the effort required to comply with the regulatory requirements. Most of the effort if focused on compliance matters, governance processes, testing and monitoring. Very little time is spent running the algorithms[144].

Some (but not all) technology companies are resisting the new laws and regulations that are being enacted and there will no doubt be some landmark battles in the next few years. Whether the financial might of the tech giants or the political will of governments wins out – we shall have to wait and see.

Finally, what about the prospects of developing a fully artificial thinking entity any time soon? A conscious, thinking, Super AI with greater capacity to learn and reason than any human?

The first question to ask is, is such a thing even possible? Let's start by proposing that the human brain is nothing more than a very complex biological machine that conforms to the general laws of nature. If that's true, then I can't think of any theoretical barrier to engineering something that's at least as good as something that blind evolution has come up with, even if we don't know how to do that today. However, if you believe that there is some spirit form, a soul or something else required for conscious intelligence, then maybe we'll never get there unless we find a way to create that "Essence of intelligence" in one form or another.

Assuming that the brain does conform to the laws of nature, then in terms of when we'll get to general AI, then that's a difficult call. However, I don't think that we'll get there using current approaches, even if we can scale up the computer power available by several orders of magnitude.

Why do I think this? Well, I don't disagree that practical applications of artificial intelligence have grown at a fantastic rate. This has resulted in many sophisticated applications with capabilities that would have been almost unimaginable just a few years ago. If you project forward these advances in a simplistic way, then it may seem that it can't be long (just a few years at most) before the capabilities of the machines exceed our wildest expectations and can outthink us in all significant ways. However, nearly all of these recent advances in artificial intelligence have been based on variants of a single approach; that is, applying massive computational power to artificial neural networks in one form or another.

Almost everyone working in artificial intelligence is focused on neural network based approaches. Only a very small amount of research effort is being spent looking at alternatives. What this means is that we could end up missing something important in the search for full artificial intelligence because not enough people are looking at the problem more broadly. I'm not saying that neural networks aren't immensely important to the development of AI, but if we only consider artificial neural networks, then we are probably missing something vitally important for developing AI further.

What we are seeing is that the applications resulting from neural network approaches are increasingly powerful when it comes to singular problem-solving tasks (Narrow AI), but we don't yet seem to be near the point where we have tools that go much beyond that. The best we have today is the idea of sticking a lot of very good narrow AI together, and then adding in some human expertise to define additional rules and over-rides. This delivers something slightly "Fatter" than a single application can provide on its own, but falls far short of the dream of general AI.

Another thing that worries me about current neural network approaches is how inefficient they are. It takes hundreds of high-end computers, running for days on end, burning millions of kilowatts of electricity to build an advanced AI app that does just one thing very well. For example, playing StarCraft, or being able to identify what emotion people are feeling from their facial expressions. A well-documented example, that illustrates this issue very well, is with object recognition systems. Show a young child a single picture of a puppy once, and they'll pretty much be able to identify any dog they see. However, an object recognition system needs to be trained with thousands and thousands of images of dogs to be as good as that.

This makes me think that we aren't doing something quite right. Or, maybe we must be missing something. A simple organic human brain can learn to do thousands of things pretty well (if not quite as well as a computer) using only about 20 watts of power and a lump of organic matter that weighs, on average, only between about 1,300 and 1,400 grams. OK, it takes us years to learn all these things, but we still have learning capabilities that the machines can't yet match. In particular, some of the biggest challenges in artificial intelligence research relate to comprehension and understanding that allows us to exhibit common sense. To take learning from the wider world and pull out things that are relevant to a completely different set of problems. We can contextualize ideas and tasks within a broader understanding of the world and how it operates. This is both in physical terms about how things interreact with each other, but equally, intellectual concepts such as ethics, religion, society and politics.

This leads me to fall into the camp that believes that there is some essence of consciousness intelligence that we've not discovered or understood yet. No amount of raw computing power is going to be enough to generate full artificial intelligence. I'm not saying that this is a soul or spirit, but that there are one or more scientific discoveries that we need to make before we really understand consciousness, what it means to be self-aware and what human intelligence actually is. Until those breakthroughs occur, then I expect we'll continue to see more developments in terms of an ever-increasing number of clever AI-based apps and robots for certain tasks, but we won't see a human equivalent (or better) general thinking machine. If those discoveries are made this year, then it may not be very long before such a machine is built. If, on the other hand, it takes decades more research, then it will be quite a long time before we cede world dominance to the computers.

As a final note, if we do ever develop full artificial intelligence, that is comparable or superior to human intelligence, then it probably won't be in the form of human-like robots or a smart assistant that sits in your pocket. Instead, it's likely to be a centralized entity. By that, I mean one or more of these intelligences will be housed in server farms somewhere. They won't be physically present where we can see them. In that way, size and energy requirements are almost unimportant. If the AI needs a giant computer complex, the size of a mountain in California or Beijing, then that's fine. It can then communicate and control devices round the world via the internet. It may interact with us via a robot body but the AI itself could be thousands of miles away. Sure, if we wanted that intelligence to travel out into the solar system or beyond, then it would need to be portable due to the time lags involved[145], but that's a problem for another day...

Appendix A. Further Reading

"A little knowledge is a dangerous thing"[146]

If you find artificial intelligence interesting and want to know more, then the following are some other books, papers and magazines that might be of interest. All are non-technical in nature; i.e. no complex math or formulas.

Ajay Agrawal, Joshua Gans, Avi Goldfarb. (2018). *Prediction Machines: The Simple Economics of Artificial Intelligence.* **Harvard Business Review Press.** Most AI applications rely on prediction. The three authors are economists rather than technical experts, who present the case for AI as a more efficient way to predict those things that are important to business, both strategically and operationally.

Margaret A. Boden. (2018). *Artificial Intelligence. A Very Short Introduction.* **Oxford University Press**. This is an academically styled book that seeks to provide a brief history and philosophical description of the developments in artificial intelligence in a concise manner.

Paul Daugherty and James Wilson. (2018). *Human + Machine: Reimaging Work in the Age of AI.* **Harvard Business Press**. This book presents a nicely optimistic view of a future in which artificial intelligence supports humans in the workplace rather than supplanting them.

Steven Finlay. (2018). *Artificial Intelligence and Machine Learning for Business: A No-Nonsense Guide to Data Driven Technologies.* **Relativistic. 3rd Edition. Relativistic.** This is one of my other books. Therefore, please take this recommendation with a pinch of salt. It covers much of the same material as this book, but focuses on business applications (I may even have "Borrowed" the odd paragraph here and there for this book – but hey, why reinvent the wheel if you don't need to?)

Hannah Fry. (2019). *Hello World: How to be Human in the Age of the Machine.* **Black Swan.** A nice, well written, non-technical introduction to the world of algorithms, machine learning and artificial intelligence. Probably the book I would recommend first to someone wanting an easy going entrée to the subject (apart from this book of course!)

Information Commissioner's Office. (2017). *Big data, artificial intelligence, machine learning and data protection.* **ICO.** If you want to know more about the implications of artificial intelligence and related technologies within the context of EU/UK data protection law (The General Data Protection Regulation or GDPR) then this provides a lot of useful information. It's available for free from the UK Information Commissioner's Office website.

Cate O'Neil. (2016). *Weapons of Math Destruction. How Big Data Increases Inequality and Threatens Democracy.* **Allen Lane.** While many authors talk about all the upsides, O'Neil presents the dark underbelly of automated decision making. In particular, she discusses the bias and discrimination that can be brought to bear through the use and misuse of automated decision-making systems. This book is a great accompaniment to "The Age of Surveillance Capitalism" by Shoshan Zuboff.

Stuart Russell. (2019). *Human Compatible: AI and the Problem of Control.* **Allen Lane.** This book looks ahead to consider the problems we might face should we ever build super-human intelligences and how we might go about protecting ourselves against them. A fine compliment to Max Tegmark's book on the possible future of AI (See below).

New Scientist. https://www.newscientist.com/ If you want a weekly digest of all that is going on in the world of science and technology, presented in an intelligent but non-technical way, then there are few publications better than this. At the moment, you can pretty much guarantee that there will be at least one article on developments in AI and machine learning every week.

Max Tegmark. (2017). *Life 3.0: Being Human in the Age of Artificial Intelligence.* **Penguin.** An interesting and provocative soiree into the potential, risks and dangers of artificial intelligence and its relationship to humanity now, in the near future and what might eventually come to pass in the millennia to come.

Shoshan Zuboff. (2019). *The Age of Surveillance Capitalism: The Fight for a Human Future at the New Frontier of Power.* **PublicAffairs.** This book really digs into the issue of privacy and control, and how organizations have used their unfettered access to our personal data to boost their bottom lines at our expense.

Appendix B. Glossary of Artificial Intelligence Terminology

"If you can't explain it simply, you don't understand it well enough."[147]

By gaining an understanding of the terms below, you should be able to hold your own when talking down and dirty with the nerds, geeks and PhDs...

Accuracy. This is a measure of how often a decision-making system get things right. It is calculated as the proportion of times the system makes a correct decision. If an object recognition system is shown 50 pictures and correctly identifies 48 of them, then it's 96% accurate (100% * 48 / 50).

Activation function. This is an equation that forms one part of a neuron. The main purpose of the activation function is to standardize the neuron's output so that it lies within a fixed range. This is so that all of the neurons in a neural network produce values that have the same scale. Most commonly, the activation function forces the output to be the range 0 - 1. Popular activation functions are the "Logistic function" and the "Hyperbolic tangent function." See https://en.wikipedia.org/wiki/Activation_function for a more comprehensive list of activation functions.

Algorithm. A set of instructions, executed one after another, to complete a given task. In the world of AI, algorithms can broadly be split into two types. Training algorithms are used to discover patterns in data and to create the models that drive most AI applications. Once a model has been created, the application of the model is itself implemented as an algorithm; i.e. the steps required to create outputs from the model and then act upon them.

Android. An intelligent robot with human-like appearance that behaves in a human-like way.

Artificial Intelligence (AI). There isn't a single universally accepted definition of what AI is. However, a reasonable working definition is: The replication of biological analytical and decision-making capabilities. Examples of AI include: digital personal assistants that can answer questions, identifying what an object is in an image, vetting job applicants and the ability to beat human players at games such as chess, poker and Go.

Autoencoder. A type of neural network used to create a transformed representations of a data set. Often, this will be for the purpose of variable reduction in a similar manner to principle component analysis (PCA). Autoencoders can also be used for clustering, and to identify unusual or odd cases (outliers). This can be useful for tasks such as fraud detection.

Big Data. Any large and varied collection of data held in computer storage. The term is usually applied to data sets that are too large and complex to be processed easily using a standard laptop/PC. Big Data technologies make use of advanced computer architectures and specialist software to facilitate the rapid processing of these data sets, often in real time. AI/Machine learning is one of the primary tools used to extract value from Big Data.

Black box. A system is considered to be black box in nature if it's difficult to determine how the inputs into the system relate to the outputs generated. Systems that utilize complex neural network and/or ensemble models are often described as being black box, whereas simple scorecard, decision tree and rule-based systems are not.

Causation. The reason why something happened. For example, if asking the question: "Why do plants grow more in the summer than in the winter" two facts are presented. (1) It's warmer in summer than in winter. (2) people take more holiday in summer. Plant growth is correlated with people taking more holiday, but not caused by it. However, plant growth is both correlated with and caused by warmer weather.

Chatbot. An app that can interact with someone in a human-like way via speech or text. Typically, a chatbot that you are likely to encounter today, is designed to provide answers to questions about a specific topic. For example, an insurer's chatbot might answer questions about the company's insurance products via their website. Most chatbots are relatively basic. They are used to automate simple, frequently asked, questions in order to improve call center efficiency. More complex or unusual questions are referred to human staff to answer.

Classification model. A predictive model that estimates the probabilities of different outcomes (events). For example, based on the pattern of pixels in an image, there is a 95% probability that the thing in the image is a cat, 4% that it's a dog and 1% that it's a fish. The prediction made by the model is that the image contains a cat, because that has the highest probability associated with it. Another example is a credit scoring model. The model predicts the probability of a customer defaulting on their loan. See also, regression model.

Classification and Regression Tree (CART), *See* decision tree.

Cluster(ing). Clusters are groups that contain people or things with similar traits. For example, people with similar ages and incomes might be in one cluster, those with similar job roles and family sizes in another. Clustering algorithms are a form of unsupervised learning (see below).

Convoluted (neural) network. A deep neural network where not all neurons in one layer are connected to all the neurons in the next layer. With fewer weights, the very considerable time required to train a network is much reduced. A classic application is object recognition. Imagine you have an image comprising 256 * 256 (65,536) pixels. A traditional neural network would need 65,536 weights for each neuron in the first layer, one for each pixel. By segmenting the image into say, 64 sub-regions each with 32 *32 pixels and only connecting neurons within each sub-region, the number of weights is reduced to just 1,024 per neuron. This works because the most important features in images are usually close together (i.e. in the same or neighboring sub-regions). Pixels at opposite sides of the image are much less important.

Correlation. One variable is correlated with another if a change in that variable occurs in tandem with a change in the other. It's very important to understand that just because two events are correlated with something does not necessarily mean that one thing causes the other. Purchases of mozzarella cheese are correlated with pepperoni sales but people don't buy pepperoni *because* they've bought mozzarella. They buy both because they are making pizza. See also, causation.

Cut-off score, *see* decision rule.

Data lake. An organizational data store. What differentiates a data lake from a data warehouse is that a warehouse is usually in a much more organized and structured format than a data lake. To put it another way, a data lake is more of a "Data dump" whereas a data warehouse usually contains data that has been cleaned and processed into a nice neat format that supports specific organizational purposes such as management reporting.

Data set. A collection of data available for analysis. Examples of data sets include a spreadsheet containing patient records, details of peoples' credit card transactions and audio recordings of customer calls to a contact centre.

Data mining. Data mining came to prominence in the 1980s. It describes the use of computers to find useful information from large quantities of data that are too big for a human to analyze easily on their own. Although data mining includes techniques found in machine learning and artificial intelligence, the term it is not widely used these days.

Data science. The name given to the skill/art of being able combine mathematical (machine learning) knowledge with data and IT skills in a pragmatic way, to deliver practical value add machine learning-based solutions.

Data scientist. The name given to someone who does data science. Good data scientists focus on delivering useful solutions that work in real-world environments. They don't get too hung up on theory. If it works it works.

Data warehouse. A structured collection of data used to support management reporting and decision-making. A data warehouse usually contains multiple data tables (like a set of Excel worksheets). A retailer may have a data warehouse that contains customer details in one table, information about the products that they sell in another and a records of customer transactions in a third. The three tables can then be used to generate reports about customer purchase behaviour.

Decision bias. One type of bias that can manifest itself in a data driven decision-making system. The system makes biased decisions because the training data used to develop it was also biased. See also, Sample Bias.

Decision rule. Models generate scores; i.e. numeric outputs. Decision rules are used to decide what action to take on the basis of the score. If the score is above a given value (the cut-off) do one thing, otherwise, if the score is below the cut-off, then do something else. For example, when assessing people for a medical condition, only those scoring above the cut-off score; i.e. those with the highest risk of developing the condition, are offered treatment.

Decision tree. A type of predictive model created using an algorithm that recursively segments a population into smaller and smaller groups. Also known as a Classification and Regression Tree or CART because they can be used for both classification and regression.

Deepfake. Something artificial, often created using machine learning techniques, that is passed off as the real thing. For example, creating a fake media clip of a well-known politician or celebrity, and inserting it into a real recording to given the appearance of them saying something that they didn't.

Deep learning. Predictive models based on complex neural networks (or related architectures), containing many artificial neurons spread across many layers. Deep leaning is proving very successful at many complex tasks such as object recognition and language translation.

Development sample. To develop a model, a data scientist will gather a sample of data (training data). The portion of the training data used to train the model is referred to as the development sample. This contrasts with the validation sample, which is used to test the accuracy of the model once training is complete. The development samples needs to contain at least several hundred cases but larger samples lead to better results. Complex AI system will use development samples that contain many millions of examples.

Ensemble. Sometimes, several different models will be developed using different methods and different parts of the training data. Each model predicts the same outcome but, because they have been constructed differently, they produce slightly different predictions. The ensemble combines all of the individual predictions together, to produce a new prediction, which is often more accurate than any of the individual predictions on their own. A simple ensemble can comprise just 3 or 4 models but complex ensembles can contain thousands of separate models.

Expert system. An expert system aims to replicate the deductive reasoning and decision-making capabilities of experts in a particular field. A typical expert system contains two main components. The first is a knowledge base that holds information about the field of interest. The second is an inference engine that, when presented with a set of facts, applies rules to the information contained in the knowledge base to try and determine a suitable answer. For example, given a set of symptoms, what disease is the patient most likely to have?

Feed forward neural network. This is a neural network where the connections are all in one direction from one layer of neurons to the next; i.e. all of the outputs from the neurons in one later provider the inputs to the neurons in the next layer. There are no backward connections or connections to other layers apart from the next one. Most neural networks are feed forward networks.

Forecast horizon. When predicting future events, the forecast horizon refers to the period of time over which the prediction applies. If I build a model to predict the likelihood of people defaulting on their credit cards, then the model will be designed to predict the likelihood of default in say, the next year. However, if I am predicting survival rates for a particular disease, then a forecast horizon of five or ten years may be more appropriate.

General Adversarial Networks (GANs). These comprise two neural networks that use the outputs from each other to learn and improve. One network is trained to create something that appears real, such as an image by a famous painter or soundbite from a celebrity. The second network is trained to identify if the items are real or fake. After many cycles of training, you have a network that can produce things that appear real, but are in fact, completely artificial.

Gains chart. There are many different ways to assess how well a predictive model performs. A gains chart is one very popular visual tool that is used to assess how good a predictive model is. Gains charts are often used in conjunction with lift charts.

Gartner Hype Cycle. New technologies tend to be over-hyped initially, then under-hyped, before eventually ending up between the two. The Gartner hype cycle captures this behaviour very elegantly. More information is available from the Gartner website: .https://www.gartner.com/technology/research/methodologies/hype-cycle.jsp

GDPR. General Data Protection Regulation. (Regulation (EU) 2016/679). This is EU data protection legislation that came into force in 2018. The GDPR provides detailed instructions as to how organizations must gather, store and process personal data in EU countries. The scope of the GDPR is the processing of all personal data but it has some specific clauses that are relevant to AI technologies. In particular, people have the right to a detailed explanation as to how an automated decision was made about them if that decision has a significant impact upon them. People also have the right to demand that important decisions are independently reviewed by a human being if they disagree with a decision made by a fully automated system.

General AI. An artificial system that can operate intelligently across many different problem domains and which can adapt and learn new skills in a similar way to a person, is described as possessing general intelligence. No AI systems in the world today can be said to possess General AI. See also, Narrow AI.

Gradient descent. A general-purpose algorithm that is widely used in optimization problems. In the context of AI/machine learning, gradient descent is used in training algorithms to determine the weights in a model. A commonly used analogy is trying to get to the lowest point in the local landscape as quickly as you can. To do this, you identify the steepest downward path and take a step along it, so moving towards the lowest point. You repeat the process until you can't get any lower. When building a model, deciding which weights to adjust and by how much, are analogous to determining which direction to go in and how big your step size should be.

Hadoop. A software solution for storing massive amounts of data distributed across multiple computers. The core Hadoop software is free (open source) and is implemented via a cluster of cheap, off the shelf, desktop PCs/servers. A Hadoop cluster is extremely scalable and far cheaper to build and run than a traditional supercomputer. If you need more storage, then you just buy a few more cheap PCs and plug them into the system. See also, MapReduce.

Imitation Game (The). The imitation game was the name Alan Turing gave to his test to determine if a machine could be classified as intelligent. Today, this is more commonly referred to as the Turning Test. In the Imitation Game / Turing Test, a machine is deemed to be intelligent if a human judge can't distinguish which is the human and which is the machine after engaging both in conversation.

Internet of things (The). Everyday devices such as cars, heating system, TVs and even light bulbs can be connected to the internet and each other. The Internet of Things (IoT) describes these types of connected devices and how they are used. For example, heating systems that predict that they will break down in a few days' time, find a slot in your smartphone diary and then arrange an engineer visit to fix the problem before it occurs. More and more devices are being seen with internet connectivity, but the IoT concept is still in its infancy.

K-Means clustering. A method of grouping cases into a set of K similar groups (clusters). This is achieved by minimizing the differences between observations in each of the clusters. This is probably the most common type of clustering in use today.

Lift chart. A graphical tool that is widely used to demonstrate how well a predictive model performs. Lift charts are often used in conjunction with gains charts.

Linear model. One of the earliest types of predictive model used in automated decision-making systems. These days, other types of model can often deliver more accurate results, but linear models remain popular because they are easy to understand and use. This means that it's easy to describe why the model delivers a given score, and hence why certain decisions were made. A popular way of representing linear models is in the form of a scorecard, such as the one introduced in Chapter 3.

Linear regression. This is one of the most popular methods for creating linear models and scorecards. Its development dates back more than 100 years.

Logistic regression. Like linear regression, this is another popular method used for creating linear models and scorecards. Logistic regression is theoretically more appropriate than linear regression for classification type problems, although in practice it delivers models that display simpler discriminatory ability in most cases.

Machine learning. Machine learning covers a range of algorithms that have been developed from research into artificial intelligence and pattern recognition. Machine learning algorithms are primarily used to find patterns (features) in data. The algorithms used to train (deep) neural networks are one example of machine learning in practice.

MapReduce. A programming approach that enables data stored on Hadoop (and other Big Data platforms) to be processed very quickly. MapReduce works by splitting data processing tasks into lots of smaller sub-tasks that can then be implemented in parallel across the network of computers that comprise a Hadoop network.

Model. A mathematical representation of a real-world system or situation. The model is used to determine how the real-world system would behave under different conditions. In AI, most models predict some type of outcome that is used to make decisions and take actions. See also, predictive model.

Monitoring. The performance of models tends to deteriorate and change over time. It is therefore prudent to instigate a monitoring regime following model implementation to measure how models are performing on an ongoing basis. Models are redeveloped when the monitoring indicates that a significant deterioration in model performance has occurred, or the model is sub-optimal in some other way. For example, if illegal or undesirable biases are detected.

Narrow AI. Artificial intelligence applications which are very good at just one or two things, but which cannot be applied beyond the problems for which they have been designed. All AI applications in use today can be described as being Narrow AI systems. See also, General AI.

Natural Language Processing (NLP). Algorithms that deal with language; i.e. speech and text processing. A primary aim of NLP is to allow computers to interact with people using everyday language. However, NLP also encompasses language analysis to identify sentiments, emotions and motives. For example, analysis of a Twitter feed to identify if the subject is being talked about in a positive or negative way, or detecting fraudulent activity from a criminal's speech patterns.

Neural network. A model constructed from a set of interconnected neurons. Neural network models are very good at capturing complex interactions and non-linearities in data in a way that is analogous to human learning. Deep neural networks (Deep learning/Deep belief networks) are very large and complex neural networks, often containing thousands or millions of artificial neurons. These types of models are favored for advanced AI tasks such as speech recognition and the navigation systems in self-driving vehicles.

Neuron. The key component of a neural network, which is often presented as being analogous to biological neurons in the human brain. In reality, a neuron is a linear model whose score is then subject to a (non-linear) transformation. A neural network can therefore be considered as a set of interconnected linear models and non-linear transformations.

Odds. A popular way to represent the likelihood of an event occurring. The odds of an event are equal to $(1/p) - 1$ where p is the probability of the event. Likewise, the probability is equal to $(1/Odds+1)$. Odds of 1:1 is the same as a probability of 0.5, odds of 2:1 a probability of 0.33, 3:1 a probability of 0.25 and so on.

Over-fitting. This is when an algorithm goes too far in its search for patterns and correlations in the data used to develop a model. The result is a model that is very accurate when measured using the training data (the development sample) used to build the model, but in practice performs very poorly when it's used to generate new outcomes using data that has not been used before.

Override rule. Sometimes, you have to carry out a certain action, regardless of the output generated by a model. A predictive model used to target people with offers for beer might predict that some children are very likely to take up the offer. An override rule is therefore put in place to prevent offers being sent to children, regardless of the score generated by the model.

Predictive analytics (PA). This term is used to describe the application of statistical or machine learning techniques to generate predictive models. There is an argument that, for all practical purposes, machine learning and predictive analytics are pretty much the same thing, given that they use the same types of data as inputs, apply the same type of algorithms and generate similar outputs (scores).

Predictive model. A predictive model is the output produced by most machine learning algorithms. The model captures the relationships (correlations) that the training algorithm has discovered. Once a predictive model has been created, it can then be applied to new situations to predict future, or otherwise unknown, events.

Predictive modelling, *see* predictive analytics.

Principle Component Analysis (PCA). PCA is a variable reduction method. This means that it's used to represent the information contained in a large number of variables with a much smaller set. The reduced set of variables are the ones that are used in the machine learning process. This helps to reduce the amount of computer power needed. PCA works on the premise that many variables are correlated with each other to a degree; i.e. contain some of the same information. Therefore, this information can be captured more efficiently using fewer variables. See also, autoencoders.

Profiling. Profiling, as defined by the General Data Protection Regulation (GDPR), is: "Any form of automated processing of personal data consisting of the use of personal data to evaluate certain personal aspects relating to a natural person, in particular to analyze or predict aspects concerning that natural person's performance at work, economic situation, health, personal preferences, interests, reliability, behavior, location or movements."[148] Profiling may or may not form part of an automated decision-making system.

Python. Python has become one of the most popular computer languages used for machine learning and the development of AI-based applications. The Python software is free and open source, with the user community able to develop new functionality and share it with other users.

R. The R language is another popular computer language for undertaking machine learning, but is also widely used for general statistical analysis. Like Python, the standard R software is free to use.

Random forest. Random forests combine together the outputs of a large number of decision trees. Each decision tree is created under a slightly different set of conditions, and hence, generate different outcomes for a given set of inputs. Random forests are one example of an ensemble model.

Recurrent neural network. A neural network model that can use previous outputs as inputs. These types of network are particularly useful where there are sequential/temporal patterns in data; i.e. the ordering of the cases in the training data is important, such as when predicting text or what is likely to happen next in a video clip.

Regression model. A model that predicts the size or magnitude of something. For example, what the temperature will be today or how long someone is expected to live. This is in contrast to a classification model which predicts the likelihood (probability) that certain events will occur. You may have a classification model to predict the likelihood of someone buying something from your store and a regression model to predict how much they are likely to spend.

Reinforcement learning. A machine learning approach where the training algorithm adjusts the weights in the model based on some measure of success resulting from an action being taken. Each time the model produces an outcome, the quality of that outcome is assessed. The training algorithm then adjusts the model weights depending on how successful it was. Reinforcement learning is viewed as more similar to the way people learn than other types of machine learning; i.e. supervised learning.

Response (choice) model. A type of classification model that is used predict the likelihood of someone choosing a given option. This type of model is used widely used in marketing to determine what products or services a customer is likely to choose (buy) so that they can be targeted with relevant offers for those products.

Sample bias (Sampling bias). This is when the training data does not contain a representative sample of the population. Common sample biases includes under-representation of certain ethnicities, genders, ages and disabilities.

Score. Each output generated by a model is usually a number (a score). For a classification model, each score represents the probability of an event occurring, e.g. the probability that someone will prove to be a good hire or the likelihood that the object in a picture is a cat. With regression models, the scores predict the magnitude of something. For example, what the temperature will be this time next week or how much disposable income someone has.

Scorecard. A popular way to present linear models that is easy for non-experts to understand. The main benefit of a scorecard is that it's additive. The model score is calculated by simply adding up the points that apply. Multiplication, division and other more complex arithmetic is not required.

Score distribution. A table or graph showing how the scores from a predictive model are distributed in relation to the thing being predicted. In credit scoring, for example, one expects to see an increasingly high proportion of bad loans with lower scores and a corresponding increasing proportion of good loans with high scores. Lift and gain charts are both ways of presenting a score distribution in a graphical form.

Sentiment analysis. Techniques that are used for extracting information about peoples' attitude towards things. For example, sentiment analysis can be used to analyze responses to a blog post to see if readers had a positive or negative view of the opinions expressed in the post. In artificial intelligence, sentiment analysis is used to extract information from text or speech that is then used to build predictive models or derive clusters.

Server farm. A large collection of computers, sometimes many tens of thousands, linked together to provide massive storage and computational capabilities. All of the large tech firms, including Google, Facebook, Microsoft, Amazon and Apple maintain multiple server farms to support the various internet-based services that they provided to users around the world.

Supervised learning. The application of machine learning where each case in the training data has an associated outcome which one wants to predict. The cases are said to be "Labeled." An example of supervised learning in target marketing is where each customer's response to marketing activity is known (they either bought the advertised product or they didn't). The model generated by the algorithm is then optimized to predict if customers will respond to marketing activity or not. In practice, most machine learning approaches are examples of supervised learning.

Support vector machine. An advanced type of non-linear model. Support vector machines have some similarities with neural networks.

TensorFlow. An open source library of deep learning algorithms for use with the Python programming language. TensorFlow was originally developed by Google. A key feature of TensorFlow is that it makes use of the processing capabilities of high end graphics cards to significantly increase the speed at which complex machine learning models (deep neural networks in particular) can be developed.

Threshold, *see* Decision rule.

Training. Training is a term used to describe the process of iteratively refining a model to improve its accuracy. See also, training algorithm.

Training algorithm. A machine learning algorithm that is used to determine the structure and/or weights in a model. The term is most widely used to describe algorithms that determine the weights for neural network models. The algorithm adjusts the weights in the network with the aim of optimizing the network's performance. The training algorithm terminates after a fixed number of iterations or when no further significant improvements in model performance are obtained.

Training data, *see* development sample.

Turing Test, *see* Imitation Game.

Unsupervised learning. The application of machine learning in situations where the training data does not contain labels (outcomes). Unsupervised algorithms typically group cases with similar characteristics (features) together. An example of unsupervised learning is an organization wanting to come up with an ad placement policy for an expensive luxury product, where no information exists about customers' purchasing history. Clustering is applied to group similar customers together based on their age, income, gender etc. The ad placement strategy is then targeted at individuals within clusters where the average income is high, rather than clusters with lower incomes.

Validation sample. An independent data set used to test the performance of a model after it has been constructed. The validation sample should be completely separate from the development sample and should not be used by the training algorithm to determine the structure of the model. Using a validation sample is important because machine learning sometimes reports over-optimistic results if its performance is measured using the data used to build it. To put it another way, if you evaluate a model using the original training data it can appear to be more predictive than it actually is. See also, over-fitting

Variable. A data item containing information about something. For example, someone's age or income, how fast a plane is flying, the temperature today and so on.

Weak AI, *see* Narrow AI.

Weighting. The process of making certain items of data feature more/less prominently in the data set used by a training algorithm to build a model. Weighting is one approach to reducing bias. Let's say that a sample of job applicants contains twice as many men as women. To create a more balanced training set, each woman's data would be given twice the weight of each man's. In effect, each woman's details are duplicated so that the number of male and female records are the same.

About the Author

Steven Finlay is a bloke whose been doing stuff with data and machine learning for a couple of decades now, and who likes to think of himself as a bit of a futurologist – although not necessarily a particularly good one.

He's done a lot of technical nerdy stuff in his time, but these days, his focus is on the bigger picture and broader issues associated with implementing new technologies. This is so that they are used successfully for the benefit of business and society. He holds a PhD in predictive modelling and is an honorary research fellow at Lancaster University in the UK.

Steve has previously been employed by one of the UK's leading banks to manage their inventory of credit risk models, has developed machine learning approaches for the UK government and worked for a number of consultancy groups and a credit reference agency. He is currently Head of Analytics for Computershare Loan Services (CLS) in the UK. He lives in Preston, Lancashire, the UK's newest city.

Steve has published a number of practically focused books about machine learning, artificial intelligence and financial services. His other books include:

- *Artificial Intelligence and Machine Learning for Business. A No-Nonsense Guide to Data Driven Technologies*. Relativistic.

- *Predictive Analytics, Data Mining and Big Data. Myths, Misconceptions and Methods*. Palgrave Macmillan.

- *Credit Scoring, Response Modeling and Insurance Rating. A Practical Guide to Forecasting Consumer Behavior.* Palgrave Macmillan.

- *Dice Role Tables.* Relativistic.

- *The Management of Consumer Credit. Theory and Practice.* Palgrave Macmillan.

- *Consumer Credit Fundamentals.* Palgrave Macmillan.

Notes

[1] Max Brooks. (2006). World War Z: An Oral History of the Zombie War. Crown.

[2] This is a version of Amara's Law, formulated by the American futurologist Roy Charles Amara (1925 – 2007). This stated that: "We tend to overestimate the effect of a technology in the short run and underestimate the effect in the long run."

[3] For example, Elon Musk was reported to have been talking about autonomous cross-country trips being available in 2017. General Motors promised us that self-driving rides would be available in 2019 and Ford was talking about 2021, which now looks unlikely. Matt McFarland. (2019). "Self-driving cars: Hype-filled decade ends on sobering note." CNN. https://edition.cnn.com/2019/12/18/tech/self-driving-cars-decade/index.html, accessed 02/01/2020.

[4] This is part of a quote attributed to Mark Zuckerberg which goes: "Move fast and break things. Unless you are breaking stuff, you are not moving fast enough."

[5] Yves Mersch. (2019). "Money and private currencies - reflections on Libra." Bank of International Settlements. https://www.bis.org/review/r190902a.htm, accessed 02/01/2020.

[6] I didn't used to think there was a difference between Nerds and Geeks but there is a body of opinion that they are actually very different. I really like the explanation by Laurie Vazquez on the Big Think website, which goes like this: "A Geek is an enthusiast of a particular topic or field. Geeks are "collection" oriented, gathering facts and mementos related to their subject of interest. They are obsessed with the newest, coolest, trendiest things that their subject has to offer. A Nerd is a studious intellectual, although again of a particular topic or field. Nerds are "achievement" oriented, and focus their efforts on acquiring knowledge and skill over trivia

and memorabilia." https://bigthink.com/laurie-vazquez/are-you-a-geek-or-a-nerd, accessed 02/01/2020.

[7] Ian McDonald. (2004). River of Gods. Simon & Schuster.

[8] Alan M. Turing. (1950). "Computing Machinery and Intelligence." Mind 49, p. 433-460.

[9] There is some debate as to what constitutes a pass and under what conditions. For example, how the judge is selected, and how many and what type of questions they can ask. In my view, the machine would need to be able to pass the majority of the time, when questioned at length about a wide range of topics by a person of at least average intelligence. A one-off fluke win would not count in my book.

[10] John R. Searle. (1980). "Minds, brains, and programs." Behavioral and Brain Sciences 3 (3), p. 417-457. Also see the arguments in the 2019 book by Christof Koch: "The Feeling of Life Itself: Why consciousness is widespread but can't be computed." And those by Roger Penrose in his books: "The Emperor's New Mind: Concerning Computers, Minds, and the Laws of Physics" and "Shadows of The Mind: A Search for the Missing Science of Consciousness."

[11] https://shop.colgate.co.uk/collections/connect-e1/products/e1-smart-electric-toothbrush?gclid=EAIaIQobChMIs8-92san4QIVQZztCh228AguEAAYASAAEgJbmPD_BwE, accessed 30/06/2019.

[12] https://news.samsung.com/global/samsungs-industry-leading-washing-machines-continue-to-garner-acclaim-for-their-disruptive-innovation, accessed 29/03/2019.

[13] Daniel Thomas. (2019). "Five tech trends shaping the beauty industry." BBC. https://www.bbc.co.uk/news/business-48369970, accessed 25/05/2019.

[14] https://www.captionbot.ai/, accessed 23/06/2019.

[15] https://openai.com/blog/musenet/ , accessed 23/06/2019.

[16] https://thispersondoesnotexist.com/, accessed 02/01/2020.

[17] https://www.bostondynamics.com/robots, accessed 15/11/2019

[18] BBC. (2019). "AI Judge to settle small claims disputes." https://www.bbc.co.uk/news/av/technology-47555567/ai-judge-to-settle-small-claims-disputes-and-other-news, accessed 06/04/2019.

[19] Timothy Revell. (2019). "AIs are really dumb. They don't even have the intelligence of a 6 month old." New Scientist 242(3233), p. 44-5.

[20] Douglas Heaven. (2019). "AI can't see things from another view." New Scientist 241(3221), p. 15.

[21] Chris Baraniuk. (2019). "One of these is a power drill." New Scientist 242(3227). New Scientist, p. 34-7.

[22] Rachel Metz. (2018). "Microsoft's neo-Nazi sexbot was a great lesson for makers of AI assistants." MIT Technology Review. https://www.technologyreview.com/s/610634/microsofts-neo-nazi-sexbot-was-a-great-lesson-for-makers-of-ai-assistants/, accessed 06/01/2020.

[23] Matt Day, Giles Turner and Natalia Drozdiak. (2019). "Amazon Workers Are Listening to What You Tell Alexa." Bloomberg. https://www.bloomberg.com/news/articles/2019-04-10/is-anyone-listening-to-you-on-alexa-a-global-team-reviews-audio, accessed 25/05/2019.

[24] https://developer.amazon.com/docs/ask-overviews/understanding-how-users-interact-with-skills.html, accessed 29/03/2019.

[25] Jane Wakefield. (2019). "Amazon gets closer to getting Alexa everywhere." BBC. https://www.bbc.co.uk/news/technology-50392077, accessed 22/11/2019.

[26] Typically, organizations monitor the performance of their credit scoring systems on a monthly or quarterly basis. They then update the system when there is evidence that a better credit score could be constructed or when new data become available. Sometimes it can be several years between updates.

[27] Bruce J. Kellison, Patrick Brockett, Seon-Hi Shin and Shihong. (2003). "A Statistical Analysis of the Relationship Between Credit History and Insurance Losses." Bureau of Business Research.

[28] Microsoft Azure. https://azure.microsoft.com/en-gb/services/machine-learning/?&wt.mc_id=AID529440_SEM

[29] Google Cloud Prediction API. https://cloud.google.com/prediction/

[30] Alexa Voice Service (AVS) https://developer.amazon.com/public/solutions/alexa/alexa-voice-service/getting-started-with-the-alexa-voice-service, accessed 30/03/2019.

[31] This is generally true, but lenders obviously should not lend to children and most won't lend to you if you don't have a job.

[32] That is not to say that many people won't all receive the same credit score, but that the route to arrive at that score, via the set of characteristics used to calculate it, may be different for everyone.

[33] Edward H. Shortliffe and Bruce G. Buchanan. (1975). "A model of inexact reasoning in medicine." Mathematical Biosciences. 23(3–4), p. 351-379.

[34] Rory Cellan-Jones. (2017). "Can Google police YouTube?" BBC http://www.bbc.co.uk/news/technology-39338009, accessed 28/06/2019.

[35] Leo Kelion. (2018). "YouTube toughens advert payment rules." BBC http://www.bbc.co.uk/news/technology-42716393, accessed 28/06/2019.

[36] Carl Sagan. (1990). "Why We Need to Understand Science." Skeptical Inquirer, Spring 1990.

[37] It's all to do with air pressure. The lower the air pressure the lower the boiling point of water. A rough rule of thumb is that the boiling point drops by one degree centigrade for each 1,000 feet (300 metres). At the top of Everest (28,029 feet) the boiling point of water is only about 72 degrees. This would not make a very good cup of tea.

[38] Steven Finlay. (2012). Credit Scoring, Response Modeling and Insurance Rating: A Practical Guide to Forecasting Consumer Behaviour. Palgrave Macmillan. 2nd Edn, p.8.

[39] Body Mass Index (BMI) is calculated as follows: Take someone's weight in kilograms and divide it by their height in metres squared. If someone is 180cm tall and weighs 80kg then their body mass index is: $80/(1.8 * 1.8) = 24.69$. In the UK, a BMI of 19-25 is considered normal. A BMI under 19 indicates a person may be underweight. 25-30 indicates that someone is likely to be overweight and more than 30 obese. BMI should only be taken as a guide, and other factors such as build, age and muscle mass are also important. Some athletes would be classified as overweight using BMI due to having more than the average amount of muscle mass.

[40] The decision tree is an artificial example. Therefore, it should not be taken too seriously.

[41] Sometimes, the process of identifying which data items are important is undertake via a preliminary step, before the remaining data items are supplied to the part of the algorithm that constructs the final model. In modern software however, the two steps of data selection and model training are often undertaken as a single process from a user perspective.

[42] This is a valid way of creating predictive models, called the Delphi method. The Delphi method is particularly useful in situations where there is little or no data available to apply a training algorithm to.

[43] Fair Isaac Corporation. (2003). "A Written Statement of Fair Isaac Corporation on Consumer Understanding and Awareness of the Credit Granting Process Before the United States Senate Committee on Banking, Housing, and Urban Affairs." Washington D.C., Fair Isaac Corporation: 24.

[44] It always amazes me that some people (including some "experts") quote an advantage of neural networks over other methods (see next chapter) is that they can learn and adapt in real-time. Thus implying that the same type of real-time learning can't be applied to other types of models such as scorecards and decision trees. This is an incorrect assumption.

[45] Some surveys report they remain the most popular types of models, despite all the media attention that more advanced type of models are receiving. For example see: Matthew Mao (2019). "Top Data Science and Machine Learning Methods Used in 2018, 2019." https://www.kdnuggets.com/2019/04/top-data-science-machine-learning-methods-2018-2019.html, accessed 06/01/2020. This survey reported that scorecards (regression) and decision trees were the two most widely used methods by data science professionals in 2018/9. However, their use had fallen slightly since the previous year's survey. The survey also reported an increase in the use of neural network based methods from previous years.

[46] Frank Herbert. (1965). Dune. Chilton Books.

[47] Rosenblatt, F. (1958). "The perceptron: A probabilistic model for information storage and organisation in the brain." Psychological Review 65.

[48] Their full title is Artificial Neural Network or ANN, but these days, most people just use the term Neural Network.

[49] Rumelhart, D. E., Hinton, G. E. and Williams, R. J. (1986). "Learning representations by back-propagating errors." Nature 323(6088).

[50] This approach can be used if the values represent a scale, but if the data represents a category, then a more common approach is to create a separate 0/1 flag for each condition. There would be one flag for employed, one flag for retired, one for home maker and so on.

[51] Forcing all the scores to lie in the same range is not absolutely essential, but is usually deemed to be good practice.

[52] OK, OK – I know I promised no formulas but I'm sure you'll forgive me if I throw just one into the mix. There are many activation functions, but a very common one is the "logistic function." This is calculated as

$1/(1+e^{-(\text{Initial Score})})$. The value of e is 2.718. e is a bit like PI (3.142) in that it appears in all sorts of places in mathematics and statistics. If the initial score was say, 4.2, then following the application of the logistic function, the final score output by the neuron would be 0.985.

[53] i.e. the output from a single neuron (and also a neural network with no second layer) is equivalent to logistic regression, which is a popular statistical method used in machine learning for finding the points associated with a scorecard.

[54] This is because standard neural networks, where all the inputs are provided to every neuron in the first layer, do not account for the spatial nature of images. For example, the pixels at the top left of an image often have no relation to the pixels at the bottom right. Therefore, it makes sense not to provide all inputs to every neuron.

[55] The models don't have to be neural networks but, in nearly all practical applications, neural network models are used due to their superior performance.

[56] There are lots of different flavours of GANs – some use random inputs as I've described, but some types of GAN are designed to transform inputs rather than create something from scratch. For example, to transform a photograph you've taken on your phone into the style of a painting by a famous artist. In this case, the inputs would be your photograph and the output would be the new image.

[57] https://thispersondoesnotexist.com/

[58] Assuming you were not infringing copyright in the first place when you used the content to train the model.

[59] Will Knight. (2019). "Amazon Has Developed an AI Fashion Designer." MIT Technology Review. https://www.technologyreview.com/s/608668/amazon-has-developed-an-ai-fashion-designer/, accessed 25/08/2019.

[60] Bernard Marr. (2019). "Artificial Intelligence Can Now Copy Your Voice: What Does That Mean For Humans?" Forbes. https://www.forbes.com/sites/bernardmarr/2019/05/06/artificial-intelligence-can-now-copy-your-voice-what-does-that-mean-for-humans/#56d2f34072a2, accessed 28/12/2019.

[61] Google Research Blog. (2016). "AlphaGo: Mastering the ancient game of Go with machine learning." https://research.googleblog.com/2016/01/alphago-mastering-ancient-game-of-go.html, accessed 30/06/2019.

[62] The shape of a protein is key to determining how it will interact with other substances. The task here is determining the shape, when given a list of the component parts of the protein. In theory, you could go through every possible configuration to find the actual shape of a given protein, but there are so many possible combinations that to fully explore them all could take trillions of years using the most powerful computers available.

[63] In many countries, it is illegal for someone who is currently bankrupt to obtain new credit until the bankruptcy is discharged. In the UK, bankrupts are usually discharged 1 year after the bankruptcy has been established.

[64] Agatha Christie. (1969). Hallowe'en Party. Collins Crime Club.

[65] There are also forms of supervised clustering. One such example is the popular K-nearest neighbour approach. When a prediction is required for a new case, the algorithm finds the K cases in the development sample that are most similar to it. The model score is calculated as the proportion of the K cases which displayed the behavior (outcome). If K=200, then the algorithm would find the 200 cases most similar to the case a prediction is required for. If say, 18 out of the 200 cases display the behavior, then the score is 0.09 (18/200). What value of K to use is usually determined by trial and error.

[66] There are analytical tools that can collapse a large number of data items into a 2 or 3-dimensional representation, which captures the most important features of those data items. Examples of these methods include principle component analysis and Self-Organizing Maps (SOMs).

[67] Experian's Mosaic classifications also exist in US, Germany and many other countries. A host of other companies offer similar clustering based customer profiling products.

[68] http://www.experian.co.uk/blogs/latest-thinking/consumer-segmentation-new-data-means-new-insight/, accessed 11/07/2019.

[69] This can require a lot of computer power. Therefore, sometimes a simpler approach is adopted, to assess which of the existing clusters the new case is most similar to, and that case is then assigned to that cluster.

[70] https://zbmath.org/about/, accessed 11/07/2019.

[71] 904,636 papers were added to the medline database in 2018. https://www.nlm.nih.gov/bsd/stats/cit_added.html, accessed 11/07/2019.

[72] In chess, each piece is assigned a value aligned to how important it is. A pawn is assigned a value of 1. A bishop is considered three times more valuable than a pawn, and therefore, has a value of 3. A rook (castle) has a

value of 5 and so on. Consequently, a simple view of the game can be determined by summing up the value of each players' pieces remaining on the board. An actual assessment of who is winning during a game is of course far more complex than this. All advanced chess programs consider the position of the pieces as being just as important, if not more so. However, a simple points based measure of success is very easy to understand and calculate.

[73] Knapton, S. and Watson, L. (2017). "Entire human chess knowledge learned and surpassed by DeepMind's AlphaZero in four hours." The Telegraph. https://www.telegraph.co.uk/science/2017/12/06/entire-human-chess-knowledge-learned-surpassed-deepminds-alphazero/, accessed 11/07/2019.

[74] Actually, you can – if you use a flight simulator rather than a real plane. This type of approach is one that a number of organizations are pursuing in developing of autonomous vehicles.

[75] Issac Asimov. (1957). The Naked Sun. Doubleday.

[76] Actually, according to the World Health Organisation, more than 20% of the world's population does not have access to clean drinking water. https://www.who.int/news-room/fact-sheets/detail/drinking-water, accessed 02/01/2020.

[77] ITU. (2018). "Measuring the Information Society Report Volume 1." ITU Publications, p.13.

[78] Harold Wilson. (1963). "Labour's Plan for Science." The Labour Party. This is the famous "White heat of technology" speech that Wilson gave at the Labour party conference in Scarbourgh, UK in 1963 – although what he actually talked about was the "White heat of this revolution."

[79] There were other important factors as well, such as globalization and the transfer of a lot of manufacturing capacity to developing nations where it was cheaper to produce things.

[80] Mark Overton. (2011). "Agricultural Revolution in England 1500 – 1850." BBC. http://www.bbc.co.uk/history/british/empire_seapower/agricultural_rev olution_01.shtml, accessed 09/08/2019.

[81] The 22% and 1% figures are only loosely comparable, given factors such as imports and exports, and changes in working patterns.

[82] Frey, C. and Osborne, M. (2013). "THE FUTURE OF EMPLOYMENT: HOW SUSCEPTIBLE ARE JOBS TO

COMPUTERISATION." Technological Forecasting and Social Change 114.

[83] The paper does not give a definitive date but states: "According to our estimate, 47 percent of total US employment is in the high risk category, meaning that associated occupations are potentially automatable over some unspecified number of years, perhaps a decade or two." Which gets us to somewhere around 2030-2035.

[84] PwC. (2017). "Up to 30% of existing UK jobs could be impacted by automation by early 2030s, but this should be offset by job gains elsewhere in economy." PWC. https://www.pwc.co.uk/press-room/press-releases/Up-to-30-percent-of-existing-UK-jobs-could-be-impacted-by-automation-by-early-2030s-but-this-should-be-offset-by-job-gains-elsewhere-in-economy.html , accessed 12/09/2019. The study made clear that the 38% figure for the US was a theoretical maximum based only on technological feasibility. It did not take into account legal, economic and social issues; i.e. just because it was technically feasible to automate a job did not mean that it would be.

[85] PwC. (2018). "UK Economic Outlook. PWC.", p.36. https://www.pwc.co.uk/economic-services/ukeo/ukeo-july18-full-report.pdf, accessed 13/09/2019.

[86] Nedelkoska, L. and G. Quintini. (2018), "Automation, skills use and training.", OECD Social, Employment and Migration Working Papers, No. 202, OECD Publishing, Paris, http://dx.doi.org/10.1787/2e2f4eea-en., accessed 06/05/2018.

[87] 2010 is rather arbitrary, but concerns about artificial intelligence were certainly being raised many years before I wrote this book. For example, see the 2011 book: "Race Against The Machine: How the Digital Revolution is Accelerating Innovation, Driving Productivity, and Irreversibly Transforming Employment and the Economy" by Erik Brynjolfsson and Andrew McAfee.

[88] Why have I used the US employment rate? Because the US is the largest economy in the world (In terms of GDP) and it's also where a significant proportion of the current wave of new AI technology is being developed and deployed. That's not to say a lot isn't going on in other countries, but the US is probably the single most representative one.

[89] United States Department of Labor. Bureau of Labor Statistics. Series LNS14000000.

[90] For example, a survey of 590 business leaders undertaken by KPMG in 2019, reported that only 14% expected to make people who were displaced by intelligent automation redundant. Most were seeking to retrain or redeploy people. KPMG. (2019). "Easing the pressure points: The State of Intelligent Automation." KPMG, p.27. https://assets.kpmg/content/dam/kpmg/xx/pdf/2019/03/easing-pressure-points-the-state-of-intelligent-automation.pdf, accessed 06/09/2019.

[91] United States Department of Labor. Bureau of Labor Statistics. NonFarm Business Sector. Labour Productivity. August 2019 press release.

[92] In the book: "Alice Through the Looking Glass", by Lewis Carrol, the Red Queen's race refers to Alice making no apparent progress to move forward, no matter however fast she ran. A general interpretation is that this is Carrol's analogy to real life, in which we exert all our efforts just maintain our relative position in society. Take your foot of the gas and you'll be left behind.

[93] Steven Finlay. (2014). Predictive Analytics, Data Mining and Big Data. Myths, Misconceptions and Methods. Palgrave Macmillan. Chapter 3.

[94] Brian O'Neill. (2019). "Failure rates for analytics, AI and big data projects =85% - Yikes!" Designing for Analytics. https://designingforanalytics.com/resources/failure-rates-for-analytics-bi-iot-and-big-data-projects-85-yikes/, accessed 15/09/2019.

[95] Sources: Current Population Survey, Bureau of Labor Statistics, and Federal Reserve Bank of Atlanta Calculations. Series: Overall 12ma. Unweighted Overall. https://www.frbatlanta.org/chcs/wage-growth-tracker.aspx?panel=2, accessed 02/01/2020. This data starts in 1997 whereas the unemployment and productivity graphs start in 1970. The reason for this is that I found it difficult to find reliable data on wages before 1998.

[96] The various surveys I've quoted in this chapter tend to agree in terms of the type of jobs and tasks that are most and least at risk of automation.

[97] Arthur C. Clark. (1968). 2001 a Space Odyssey. Hutchinson.

[98] For more discussion about the issue of incorrect problem specification leading to problems, see Stuart Russell's book "Human compatible. AI and the Problem of Control." (Chapter 5).

[99] Ian Banks published several books in the Culture series, the first of which was: Consider Phlebas: A Culture Novel (The Culture) in 1988.

[100] Steven Finlay. (2017). "We Should Be as Scared of Artificial Intelligence as Elon Musk Is." Fortune. https://fortune.com/2017/08/18/elon-musk-artificial-intelligence-risk/, accessed 18/11/2019.

[101] BBC News. (2019). "China due to introduce face scans for mobile users." BBC News. https://www.bbc.co.uk/news/world-asia-china-50587098, accessed 22/12/2019.

[102] Leo Kelion. (2018). "Facebook gives users trustworthiness score." BBC News. https://www.bbc.co.uk/news/technology-45257894, accessed 23/12/2019.

[103] When I say appropriately, this includes things such as people not being forced into actions just because someone else says it's good for them. For example, the right not be penalized if they subsequently don't follow the health advice that they are recommended.

[104] Of course, a lot of junk mail, AKA spam, has now moved online.

[105] Steven Finlay. (2006). "Predictive models of expenditure and over-indebtedness for assessing the affordability of new consumer credit applications." Journal of the Operational Research Society. 57(6), p. 655-669.

[106] See the book, "The Age of Surveillance Capitalism: The Fight for a Human Future at the New Frontier of Power" by Shoshana Zuboff for a detailed discussion about the problems of privacy and surveillance that has evolved as a consequence of tech companies having almost unlimited access to data about us.

[107] Robert Booty. (2019). "Benefits system automation could plunge claimants deeper into poverty." The Guardian. https://www.theguardian.com/technology/2019/oct/14/fears-rise-in-benefits-system-automation-could-plunge-claimants-deeper-into-poverty, accessed 26/10/2019.

[108] Luke Rhinehart. (1956). The Dice Man. William Morrow. In the book Luke, a psychiatrist, advocates a philosophy of making all important life decisions via the role of the dice.

[109] Somon Maybin (2016). "How maths can get you locked up." BBC News. https://www.bbc.co.uk/news/magazine-37658374, accessed 03/12/2019.

[110] For example, by copying the data about each older person, so that it features more than once in the training data.

[111] This may well lead to a less accurate model overall, but the accuracy would more likely be equally good for young and old people alike.

[112] Matthew Wall. (2019). "Biased and wrong? Facial recognition tech in the dock." BBC News. https://www.bbc.co.uk/news/business-48842750, accessed 01/12/2019.

[113] Adam Vaughan. (2019). "UK launched passport photo checker it knew would fail with dark skin." New Scientist. https://www.newscientist.com/article/2219284-uk-launched-passport-photo-checker-it-knew-would-fail-with-dark-skin/, accessed 28/10/2019.

[114] This example is adapted from: Steven Finlay. (2014). Predictive Analytics, Data Mining and Big Data. Myths, Misconceptions and Methods. Palgrave Macmillan, p. 92.

[115] The Articles of Association describe what the company's aims are; i.e. what the company is set up to do. Often, this includes maximizing the return for shareholders and owners, but can include other objectives such as supporting charities or engaging in community action.

[116] For example, they have less travel options than better off people, and hence, have more limited choice in terms of the stores they can shop in. They may also have financial problems, which means shopping on-line is difficult because they don't have access to easy credit (i.e. a credit card) and so on.

[117] OECD. (2018). "Personalised Pricing in the Digital Era." OECD, p. 24-5.

[118] Michael J. Rosenfeld, Reuben J. Thomas, and Sonia Hausen. (2019). "Disintermediating your friends: How online dating in the United States displaces other ways of meeting." Proceedings of the National Academy of Sciences of the United States of America. 116(36).

[119] The precise model algorithms used will differ by purpose but the training process employed to develop the models will be similar. The key difference is the different training data used and what the model is tasked to predict. Therefore, one tends to see the same general principles being applied in terms of how recommendations are generated by the resulting models, regardless of the type of product or service.

[120] Issac Asimov. (1950). I Robot. Gnome Press.

[121] The formulation of utilitarianism is attributed to the British philosophers Jeremy Bentham (1748 - 1832) and John Stuart Mill (1806 – 73).

[122] Kant termed this the "Categorical Imperative."

[123] Google used to promote "Do no evil" amongst its employees, but have sought to replace this with alternative statements in recent years.

[124] John Locke (1632-1714) is credited with formulating the concept of everyone having the right to life, liberty and property for all. Ideas that were later to be incorporated into the American Declaration of Independence and the French Declaration of the Rights of Man.

[125] Steven Finlay. (2014). Predictive Analytics, Data Mining and Big Data. Myths, Misconceptions and Methods. Palgrave Macmillan, p. 87-9.

[126] Paul Finlay. (2000). Strategic Management. An Introduction to Business and Corporate Strategy. Pearson Education Limited, p.75.

[127] The Equal Credit Opportunity Act (ECOA) was passed to prevent unfair discrimination against certain groups when applying for credit. As a result, credit granting decisions cannot be made wholly or partly on the grounds of race, religion, gender, nation of origin or marital status. Age can be used to a degree, but not in a negative way; i.e. to decline a credit application.

[128] European Council of Europe. 1981. "Treaty 108. Convention for the Protection of Individuals with regard to Automatic Processing of Personal Data."

[129] The GDPR (Article 9) defines this type of data as "special category data." This includes personal data revealing racial or ethnic origin, political opinions, religious or philosophical beliefs; trade union membership; genetic data; biometric data (where used for identification purposes); data concerning health, and data concerning a person's sex life or their sexual orientation. There are a number of situations where this type of data can be used, for example where an individual has intentionally made the information public.

[130] For example, the Consumer Privacy Act (AB 375) that came into force in California on 01/01/2020, which has some similar provisions to those found in the GDPR.

[131] I searched the entire text of the GDPR, and the associated "recitals" that provide further explanation of each article, but there was no mention of the words bias or biased anywhere in the entire regulation. Instead the text uses terms such as "fair" and "transparent."

[132] GDPR Article 5(1) requires personal data to be processed lawfully, fairly and in a transparent manner. See also Recital 60.

[133] GDPR Article 22 grants individuals the right not to have decisions about them made using solely automated means, where those decisions have a legal or similarly significant impact upon them.

[134] GDPR Articles 13 and 14.

[135] ICO. (2019). "ICO and The Alan Turing Institute open consultation on first piece of AI guidance." https://ico.org.uk/about-the-ico/news-and-events/news-and-blogs/2019/12/ico-and-the-alan-turing-institute-open-consultation-on-first-piece-of-ai-guidance/, accessed 07/12/2019.

[136] The "Three laws of robotics" that Asimov put forward in his Robot novels where: (1) "A robot may not injure a human being or, through inaction, allow a human being to come to harm." (2) "A robot must obey orders given it by human beings except where such orders would conflict with the First Law." And (3) "A robot must protect its own existence as long as such protection does not conflict with the First or Second Law."

[137] For example, see the following meta-study: Marc Orlitzky, Frank Schmidt and Sara Rynes. (2003). 'Corporate Social and Financial Performance: A Meta-analysis.' Organization Studies, volume 24, number 3, p. 403-441.

[138] Attributed to the author William Gibson.

[139] Harold Wilson. (1963). "Labour's Plan for Science." Speech to the Labour Party conference in 1963. The Labour Party.

[140] World Health Organization. (2019). https://www.who.int/news-room/fact-sheets/detail/sanitation, accessed 02/01/2020. WHO reports that 2 billion people (of 7.8 billion) in the world do not have access to basic sanitation facilities such as toilets or latrines. That's more than a quarter.

[141] Caroline Frost. (2017). "Tomorrow's World' Introduced These Life-Changing Inventions - Can You Guess The Year?" Huffpost. https://www.huffingtonpost.co.uk/entry/tomorrows-world-inventions_uk_5909ed76e4b02655f842ee44, accessed 02/01/2020.

[142] The Guardian. (2018). "Amazon ditched AI recruiting tool that favored men for technical jobs." The Guardian. https://www.theguardian.com/technology/2018/oct/10/amazon-hiring-ai-gender-bias-recruiting-engine, accessed 04/01/2020.

[143] In the UK, the maximum fine under data protection legislation prior to the introduction of the GDPR was £500,000.

[144] For models used to calculate a bank's capital requirements (cash to prevent the bank going bust in recessionary times), then these models must be approved as fit for purpose by an independent third party and detailed checklists produced to show that the calculations conform with all relevant regulations. The models are then passed to regulators for final approval, which requires detailed and comprehensive documentation to be provided

to explain exactly how the algorithms have been developed. Running the algorithms to produce the models is perhaps no more than 4-5% of the overall time spent on the project.

[145] The time to send a message to the nearest planets in our solar system (Venus and Mars) averages about 10 minutes. For the other planets, it can take several hours. Therefore, it would not be possible for an intelligence based on earth to control complex fast moving processes directly.

[146] Alexander Pope. (1711). An Essay on Criticism.

[147] Attributed to Albert Einstein.

[148] GDPR Article 4.

Made in the USA
Columbia, SC
02 October 2020